Black Space

Black Space

The Nazi Superweapons That Launched Humanity Into Orbit

DAVID AXE

Pen & Sword
MILITARY
AN IMPRINT OF PEN & SWORD BOOKS LTD.
YORKSHIRE – PHILADELPHIA

First published in Great Britain in 2023 by
PEN AND SWORD MILITARY
An imprint of
Pen & Sword Books Limited
Yorkshire – Philadelphia

ISBN 978 1 39901 423 6

A CIP catalogue record for this book is available from the British Library.

Typeset in Times New Roman 12/16 by
SJmagic DESIGN SERVICES, India.
Printed and bound in the UK by CPI Group (UK) Ltd.

Pen & Sword Books Limited incorporates the imprints of Atlas, Archaeology,
Aviation, Discovery, Family History, Fiction, History, Maritime, Military, Military
Classics, Politics, Select, Transport, True Crime, Air World, Frontline Publishing,
Leo Cooper, Remember When, Seaforth Publishing, The Praetorian Press,
Wharncliffe Local History, Wharncliffe Transport, Wharncliffe True Crime and
White Owl.

For a complete list of Pen & Sword titles please contact
PEN & SWORD BOOKS LIMITED
47 Church Street, Barnsley, South Yorkshire S70 2AS, United Kingdom
E-mail: enquiries@pen-and-sword.co.uk
Website: www.pen-and-sword.co.uk

Or
PEN AND SWORD BOOKS
1950 Lawrence Rd, Havertown, PA 19083, USA
E-mail: Uspen-and-sword@casematepublishers.com
Website: www.penandswordbooks.com

MIX
Paper | Supporting responsible forestry
FSC® C013604

Contents

PART THREE: VERY HIGH

Space-planes, space stations,
one very bad movie … and the Moon

PART FOUR: BEYOND JUPITER

A very good rocket, a very bad rocket and
a prayer for two moonwalkers

Introduction

We've never fought a war in space.

Yes, the world's great powers deploy satellites that aid the wars they fight on the ground, plus a few satellites whose main purpose is to tamper with or even destroy *other* satellites. And yes, they possess civilization-ending nuclear weapons that travel *through* space. But there are no weapons in orbit, no orbital fortresses poised to fry entire cities with no warning. No country deploys bombers that take off from Earth, punch through the thin border between the atmosphere and vacuum and take advantage of that lofty altitude to speed across the globe on missions of mass destruction. This despite the best efforts of US president Ronald Reagan and his Strategic Defense Initiative.

All these systems were under serious consideration at one time or another in the seven decades between humanity's first space mission and today. And no wonder, the era of frantic, dueling, American and Soviet space-exploration efforts – which stretched from the end of the Second World War to the United States' successful Moon landing in July 1969 – had its roots in Nazi Germany.

Its top scientists, most notably rocket-designer Wernher von Braun, were at best close associates of Nazis. Its basic technologies were derived from Nazi superweapons, in particular von Braun's V-2 rocket. The Nazis' slaves – Slavic prisoners of war, Jews and other 'undesirables' – built much of that hardware and died doing so. The driving concepts in early space-exploration were, fundamentally, genocidal.

Space could have become a battleground – and a terrible one, at that, considering that Nazi ideas and technology, not to mention actual Nazis, boosted humanity into orbit. Remarkably, as of the third decade of the twenty-first century, Earth's orbit overwhelmingly remains a place for cooperation, exploration, and peace, even with Russia, China and the United States deploying so-called 'dual use' satellites that could attack other spacecraft.

It's possible to triangulate the inflection point, the span of time where trends in politics, science and the lives of powerful men turned; the moment, years actually, where the seemingly inevitable war in space gradually became … evitable. The pivot point arguably was 29 July, 1958, the day US president Dwight Eisenhower reluctantly signed the law creating the National Aeronautics and Space Administration. The US military gradually ceded to NASA, a civilian agency, leaders of American efforts in space.

That pivotal moment, the decades leading to it and the years that followed all are subjects of this brief book. Not coincidentally, that historical arc maps neatly onto von Braun's own life. For von Braun more than any other single person was responsible for the ideas and technologies that made war in space possible, even – for a time – seemingly likely. For he, perhaps more than anyone, imagined the vacuum surrounding Earth as a battlefield and when history turned, and the space race became a largely peaceful contest, it was in part because von Braun – ever the opportunist – put his fingers to the solar winds, so to speak, and went along.

We've never fought a war in space. Which is not to say we never will. Chalk up our good fortune to luck or the delicate better angels of our nature, if you must but appreciate the critical role one ex-Nazi rocket-scientist played as he followed his own galactic whims in his brief, 65-year jaunt across space and time.

David Axe
Columbia, South Carolina
July 2021

PART ONE

PAPERCLIP

Science fiction, a boy stargazer,
a world war and a global scramble
to hire out-of-work Nazis

Chapter 1

Young Wernher loved rockets

Young Wernher loved rockets. It's less clear whether he cared how you made them or what you did with them and it's not clear whether he minded people dying for them or from them. Born in Wirsitz, East Prussia, on March 23, 1912 Wernher von Braun was one of three sons of an aristocrat who became secretary of agriculture in Germany's post-First World War Weimar government. He could have followed in his father's footsteps, overseeing the family's lands, perhaps entering government in some senior position. Instead, he looked first into books … and then the sky.

Von Braun voraciously read science fiction from such authors as Kurd Lasswitz, H.G. Wells and Jules Verne, but it was a work of non-fiction, Hermann Oberth's 1923 book *The Rocket into Planetary Space*, that truly shaped von Braun's young life. Oberth described, for the first time, a permanent human habitat orbiting Earth. Rockets, propelled by potent liquid fuel, would shuttle people and supplies to the station, which would spin in order to create the impression of gravity.

Not coincidentally, liquid-fuel rockets and space stations would be von Braun's twin obsessions and, as history repeatedly would bear out, young Wernher also internalized the apocalyptic implications of Oberth's fantasy. The space station Oberth posited included a huge mirror, hundreds of feet across, that would focus the Sun's light onto targeted points on Earth. Oberth described only peaceful uses for the giant space mirror – illuminating a city at night, say, or melting Arctic ice to open a winter passage for ships but he also hinted at more sinister applications.

'My space mirror is like the hand mirrors that schoolboys use to flash circles of sunlight on the ceiling of their classroom,' Oberth wrote. 'A sudden beam flashed on the teacher's face may bring unpleasant reactions.'

It's not hard to imagine how a huge orbital mirror, concentrating the energy of a star, might become a weapon. Focus a little too much light a little too long in slightly too small an area, and the mirror essentially would become a vast, city-destroying laser cannon.

Depictions by early twentieth century Germans of mankind's future in space almost always had this dual quality. On one side, they described utopia. On the other, apocalypse. Light and dark. Peace and war. This duality would come to define von Braun and, by extension, the whole space enterprise that he helped to sell to armies, governments, industry and a star-struck public. Luckily for humanity, the light eclipsed the dark. Well, *so far,* but it all might have gone very differently.

At the age of 13, von Braun received a telescope as a gift. The following year, while a student at the Ettersburg school near Weimar, he conducted his first simple experiment with crude hobbyist's rockets. He enrolled at the Berlin Institute of Technology, where he studied the new science of rocketry and in 1928, he joined the nascent German Society for Space Travel. His first mentor in the group was one of its founders, Willy Ley, the author of the popular book *The Journey into Space*.

It was an eventful year for Oberth, then president of the society. He consulted on German director Fritz Lang's science-fiction film, *Woman in the Moon*. The film's script, adapted from Ley's book *The Possibility of Interplanetary Travel*, introduced the reverse countdown – three, two, one, lift-off! – that would become a signature of space exploration all over the world.

Hermann Oberth's 1923 book *The Rocket into Planetary Space* described, for the first time, a permanent human habitat orbiting Earth. Rockets, propelled by potent liquid fuel, would shuttle people and supplies to the station, which would spin in order to create the impression of gravity. Oberth didn't specifically posit his station as an armed fortress, but reading between the lines, the station's apocalyptic potential was obvious. (Smithsonian National Air & Space Museum photo)

To promote the movie, Oberth built a small rocket. He only ever static-fired it – igniting the engine while tethering the rocket to the ground, but even that flightless demonstration deeply impressed 16-year-old von Braun. Two years later, in 1930, he became Oberth's assistant and that same year wrote a short story describing a trip to Oberth's space station. "Lunetta" didn't have much of a plot and von Braun never published it but his early scientific writing was more successful. In 1932, at the age of 20, he published an article entitled 'The Secret of the Liquid-Fuel Rocket'. It, too, bore Oberth's signature.

Hermann Oberth consulted on German director Fritz Lang's science-fiction film, *Woman in the Moon*. The script introduced the reverse countdown – three, two, one, lift-off! – that has become a signature of space missions ever since. In consulting on the movie, Oberth started a tradition of sorts. Space-travel and cinema for decades would inspire each other. Wernher von Braun more than his mentors and colleagues appreciated this synergy. Von Braun even helped to produce a big-budget biopic based on his own life, *I Aim at the Stars,* which flopped on its 1960 release. (Universum Film AG capture)

Chapter 2

Into the Third Reich

Von Braun's Oberth-inspired work on liquid rocket fuels proved to be his entrée into the Third Reich. German army officers Karl Becker and Walter Dornberger – respectively the future heads of the V-2 rocket program and the German army's weapons-research department – were mostly interested in solid-fuel rockets. Nevertheless they were impressed with von Braun, even after a liquid-fuel rocket the young scientist was working on failed during launch.

The same year von Braun's article on liquid-fuel rockets appeared in print, Becker offered him a job working on similar rocket fuels for the German army. The military paid for his research. 'To me, the army's money was the only hope for big progress toward space travel,' von Braun later said.

Von Braun attached himself to the army just in time. In 1933, Germany's many independent rocketry groups began disappearing, apparent victims of the country's march toward authoritarianism. Becker himself reportedly helped to lead this crackdown. From then until the Third Reich's destruction in 1945, the military, and only the military, would possess the power to hurl a machine into space.

On the strength of his military-funded scholarship, von Braun received his doctorate in 1934. By then he was a member of a *Schutzstaffel* horse-riding school and even had applied for membership in the group. The *Schutzstaffel*, today known by its abbreviation SS, was the Nazi Party's paramilitary wing. Von Braun quit the SS after a year, but it was a temporary schism.

Hermann Oberth, at center in profile, stands with his liquid-fuel rocket demonstrator in Berlin in 1930. Wernher von Braun, then 18 years old, stands second from right. That same year, von Braun wrote a short story describing a trip to Oberth's fictional space station. (Smithsonian National Air & Space Museum photo)

Wernher von Braun with rocket-engineer Rudolf Nebel some time in the late 1920s or early 1930s. Von Braun's work on liquid rocket fuels proved to be his entree into the Third Reich. (Smithsonian National Air & Space Museum photo)

In the mid-1930s, Wernher von Braun was a member of a *Schutzstaffel* riding school. On the Eastern Front later in the war, SS cavalry units massacred civilians behind advancing German army units. The massacres prompted a German government investigation in the early 1960s. (German Federal Archive photo)

Chapter 3

Max and Moritz

In 1933 an attempt by von Braun and his engineer Heinrich Grünow – another recruit from the German Society for Space Travel – to launch a liquid-fuel rocket ended in yet another failure at the army's Kummersdorf proving grounds, south of Berlin.

Von Braun and Grünow got to work tweaking their rocket design. They moved the stabilizing flywheel and added a valve. They produced two copies of this upgraded rocket, nicknaming them 'Max' and 'Moritz' after popular comic-book characters. Ground testing wrapped in October 1934. In December, von Braun and Grünow conducted back-to-back successful launches of Max and Moritz from a facility on the North Sea island of Borkum. Both rockets climbed to an altitude of around 6,500 feet.

That same year, Adolf Hitler's Nazi Party won 37 per cent of the vote in the national elections. Hitler didn't yet control the German government, but he was well on his way. His brownshirt thugs numbered nearly half a million and with the economy in collapse and politics in turmoil, Hitler's popularity was growing.

In 1937 Arthur Rudolph, a rocket scientist who had been working under von Braun for several years, led an effort to establish a secret military laboratory in Peenemünde on the Baltic coast. There, von Braun and his growing organization began work on the *Vergeltungswaffe Zwei* – 'Vengeance Weapon Two' – a liquid-fuel rocket that eventually became the basis for the first space-launch vehicles in the United States and the Soviet Union and, by extension, the basis of the world's first nuclear-armed intercontinental ballistic missiles.

The 13.5-ton ethanol-fueled, gyroscope-guided V-2 rocket, just shy of 46 feet from nose to tail, would carry a 2,200-pound explosive warhead a distance of 200 miles at a top speed in excess of 3,500 miles per hour. The engine fired for around a minute then an accelerometer cut off the engine when the rocket reached its design speed. Separate sets of vanes channeled the air to control roll and deflected the exhaust for steering. Gyroscopes in the nose connected to an internal guidance system that tipped the rocket over so it re-entered the atmosphere at a 45-degree angle. Because it was supersonic, the rocket struck the ground before the sound it made reached the ears of its victims.

For people in the V-2's target zones, that was a big part of the V-2's special wickedness. One instant, peace and quiet, the next instant, the world erupted. No warning, no time to hide, every moment was both safe and fatal. To the people it threatened with sudden explosive death, the V-2 was something out of a technological nightmare, but to those who designed and produced the rocket, it was clunky, temperamental clockwork – complex, unpredictable, unreliable. Actually, completing the V-2's design and fully testing and deploying it was a massive and expensive undertaking. Von Braun couldn't hope to toil away in happy obscurity, nor is it at all clear that he wanted obscurity.

In November 1937, von Braun officially joined the Nazi Party. Not joining 'would have meant that I would have had to abandon the work of my life,' he later insisted. But it's not that von Braun opposed Nazi rule, quite the contrary. In a 1952 issue of *The American Magazine*, he admitted that he 'fared relatively rather well under totalitarianism'.

Hitler first visited von Braun and his team in 1939. The scientist came away impressed with Hitler's intellect, according to von Braun's biographer Michael Neufeld. Hitler, however, was less impressed with von Braun. Hitler was ambivalent, and in the following months the rocketeers realized they'd lost the support of the Nazi Party. The Peenemünde lab withered for a lack of funding and personnel. Von

Braun feared the government would label the lab 'non-essential', a designation that would have meant its end.

However, the SS saved von Braun. In 1940, SS officials approached the scientist with the offer of a commission. Von Braun asked his

Newly-elected German chancellor Adolf Hitler watches a rocket engine demonstration at the Kummersdorf test site on 21 September, 1933. With the economy in collapse and politics in turmoil, Hitler's popularity was growing. (German Federal Archive photo)

In November 1937, Wernher von Braun officially joined the Nazi Party. Not joining 'would have meant that I would have had to abandon the work of my life,' von Braun later insisted. In 1940, Walter Dornberger urged von Braun also to join the SS, the Nazi Party's paramilitary wing. In this photo from Peenemünde in March 1941, Dornberger is in the front row at far left. Wernher von Braun is dressed in civilian clothes. (German Federal Archive photo)

mentor Dornberger for advice. Dornberger was blunt: if von Braun didn't join the SS, his career was over. So the scientist accepted the offer.

It was a happy partnership … until it wasn't.

Chapter 4

Testing the V-2

Von Braun and his team tested the V-2 twice in the summer of 1942. The first test on 13 June ended in failure when the rocket pitched out of control and crashed. The second test, on August 16, also failed. Then came a breakthrough. On October 3, 1942, a V-2 blasted off from Peenemünde. It climbed 53 miles high, becoming the first manmade object to cross that threshold that separates Earth's atmosphere from space, a hazy boundary, 62 miles up, that scientists later named the 'Karman Line' after Hungarian physicist Theodore von Karman.

The rocket nosed over and impacted in its designated target zone 120 miles away. Two years later, the besieged Nazi regime would aim thousands of V-2s at Allied troops and cities.

Hitler might have given von Braun the cold shoulder back in 1939 but a second meeting in August 1941 at Rastenburg in East Prussia, the site of Hiter's 'Wolf Lair' headquarters, was warmer. Albert Speer, the minister of armaments and war production, was a big believer in the V-2 effort and had been bringing Hitler around to that same position. By 1943, the *Fuhrer* was an unabashed fan. He bestowed von Braun with the official title of 'professor', even though the scientist didn't actually hold any academic post. The third meeting between the two men in July 1943 was, in von Braun's own recounting, a turning point in his life and work.

The timing was critical. In mid-1943, the V-2 program was teetering on the brink of collapse. Design, production and testing were spread out across multiple sites, with overlapping and at times conflicting oversight from the army, the SS and top Reich officials. The rockets von Braun's team had managed to assemble simply didn't work very well. Yes, the team had managed to successfully launch a V-2 back in October, but other tests ended prematurely when the rockets failed shortly after launch or disintegrated during re-entry as they blasted through the friction of the upper atmosphere. 'Airbursts', the testers called them.

The airburst problem would endure even after the V-2 entered operational use. For Allied troops and civilians, that was a good thing. Just over half of the V-2s Germany fired at Allied countries struck anywhere near their intended targets. Hundreds broke up in mid-air, in effect shooting themselves down.

Sabotage also contributed to the V-2's unreliability. Von Braun's workforce had long included prisoners. As the conflict wore on and the German workforce came under greater and greater strain, the proportion of slave labor in the V-2 effort only increased. They were not happy, willing workers. More than a few seized every opportunity to introduce small errors in their work – parts were the wrong size, fittings were loose.

Von Braun of course denied ever observing sabotage. 'He was almost certainly lying,' his biographer Neufeld wrote. The lie served a purpose, however, Von Braun didn't want to be known for overseeing a disloyal, undisciplined workforce.

Whoever or whatever was to blame, the V-2 program was in trouble when Speer arranged the meeting with Hitler in the summer of '43. Von Braun knew he had to make a good impression. His rocket – indeed, his professional future – depended on it. Dornberger and von Braun put together an elaborate presentation on the V-2 including scale models and a silent film of that first successful V-2 test-launch the previous October, and carefully packed everything into an He-111 bomber-transport at the Peenemünde West airfield. The He-111 deposited Dornberger and von Braun at the Wolf's Lair.

The film wasn't just some rough assemblage of test footage. It was *cinema* and von Braun's first foray into filmmaking. In a few years, the movies would turn von Braun into a star, and give him a vehicle both to reach millions with his space gospel and to escape his Nazi past. The V-2 movie von Braun produced for Hitler was surprisingly sophisticated for a first-time filmmaker. There were establishing shots of test officials. The editor cut close-ups of switches and operators into the master shot of the rocket blasting off. Following the launch, the film cut to a graphic depicting electronic tracking of the fast-flying rocket. For a denouement, the movie included archival footage of happy workers at pre-war Peenemünde. Titles flashed – *Wir haben es doch geschafft!* 'We did it after all!'

The first successful launch of a V-2 rocket, from Peenemünde on Germany's Baltic Sea coast, on 3 October, 1942. It climbed 53 miles high, becoming the first manmade object to cross that hazy threshold that separates Earth's atmosphere from 'space'. The rocket nosed over and impacted in its designated target zone 120 miles away. (German Federal Archives photo)

Hitler was impressed. 'I think we have something quite formidable here,' he said. Hitler mused that V-2 strikes on England would give Germans a morale boost. The rocket might even prove decisive in the war effort. The *Führer*'s mind was made up. He ordered Speer to give von Braun whatever labor and materials his work required. Hitler proposed that all the workers be German, but von Braun and Rostenberger sensed the German labor force was already overstretched.

At the SS's recommendation, the government approved Rudolph, still overseeing the facilities at Peenemünde, to 'recruit' laborers from nearby concentration camps.

Chapter 5

Peenemünde

Ultimately, tens of thousands of slave laborers worked on the V-2 at no fewer than three separate sites. The SS set up a camp at Peenemünde to house the slaves working at that facility. It's inconceivable that von Braun wasn't frequently in close contact with the prisoners. It's equally inconceivable that he wasn't aware of the depth of their suffering.

In August 1943, the Royal Air Force bombed Peenemünde. It's unclear exactly how British intelligence discovered the rocket work going on at the coastal base. Polish informants may have tipped off the Brits and aerial surveillance might also have played a role, but it's also possible the first major clue came from a German prisoner at a secret US military facility in Virginia – a facility that von Braun, years later, briefly would call home.

The facility in Fairfax County outside Washington, DC technically was part of the Army's Fort Hunt, which straddled former farmland that once belonged to George Washington, America's first president. The fortified prison camp and interrogation center, named for its postal address PO Box 1142, for decades was known only to staff and residents. In the early 2000s, researchers began to piece together PO Box 1142's secret history. In 2007, veterans of the facility gathered to remember their work – and to denounce brutal torture tactics that the administration of President George W. Bush practiced in Iraq.

Interrogations at PO Box 1142 were subtle, gentle, even. Agents – some of them German-speaking Jewish refugees – played games with captured Germans, befriended them, looked after their needs, fed them steak. 'We got more information out of a German general with a game of chess or ping-pong than they do today, with their torture,' Henry Kolm, a physicist turned interrogator at PO Box 1142, told *The Washington Post*.

Other methods were less convivial. Staff bugged prisoners' cells. One effective tactic was to tell a prisoner that, if he didn't cooperate, the Americans would hand him over to the Soviets. If a prisoner called that bluff, agents at PO Box 1142 would dress in Soviet uniforms … and make sure they were seen. Sometime in 1943, apparently, PO Box 1142 interrogator Werner Moritz and other agents were listening in on prisoners' conversations when the Germans began talking about rockets.

The 1940s eavesdropping technology was crude and the sound quality was poor. 'Those prisoners laid back and they just talked and it reverberates from the ceiling and it doesn't come through very well,' Moritz recalled decades later. 'Their voices don't come through very well. It was hard work. We sat there sometimes arguing with one another, 'What is this word? What is it? Where is it located?' Moritz and his colleagues eventually determined that the prisoners were talking about a rocket-factory on the Baltic coast. This conversation might have helped lead British intelligence to Peenemünde – and to the V-2 program.

In any event, many of the 600 RAF bombers that struck Peenemünde on 17 and 18 August, 1943 missed their targets and instead struck the adjacent slave camp. Hundreds died. In the aftermath of the raid, von Braun himself sifted through the flaming wreckage, trying to save the most important documentation. Peenemünde was out of commission. It would later come to light that some of the laborers at the facility had taken advantage of the chaos of the bombing to sabotage important equipment.

Eager to resume work on Hitler's favorite rocket, SS chief Heinrich Himmler approved the V-2 program to move into a new underground site in central Germany called Mittelwerk where there were two main tunnels, each more than a mile long, plus 47 smaller connecting tunnels. The SS pulled – among others – Russian, Ukrainian, Polish, Czech, French and Italian prisoners from the Buchenwald concentration camp and other prisons both to build Mittelwerk and, later, to operate its underground plants. They lived in a new camp called Mittelbau-Dora, all 16,000 of them. It took four months to prepare the tunnels and move equipment into the new underground factory.

Von Braun kept his office at Peenemünde for months after its near-total destruction by British bombers, but he observed slaves working on, and later in, Mittelwerk. 'The working conditions there were absolutely horrible,' he said. 'I saw the Mittelwerk several times, once while these prisoners were blasting new tunnels in there and it was a pretty hellish environment. I'd never been in a mine before, but it was clearly worse than a mine.'

Andre Sellier, son of a French trade-union leader, was one of the prisoners who toiled away in this 'hellish environment'. Writing about his experiences later in life, Sellier recalled von Braun traveling to Mittelwerk 'a great deal' in late 1943. 'That time frame coincided with the most painful period for the prisoners.'

The prisoners' camp was badly overcrowded. Everyone had lice. In one account that Sellier obtained, a prisoner had to throw away a prized wool sweater because 'there were almost as many lice as there were stitches'.

The tunnels themselves were dark, cold and damp. Mud in and around the underground work sites was often feet deep. During the initial excavations, the Germans allowed the workers just a few hours of sleep between shifts. There was no time to bathe, so filthy miners collapsed onto their mattresses, which quickly became 'mud pies', one prisoner told Sellier. The food was terrible and in short supply. Shoes wore out and the prison administrators were slow to replace them. Prisoners could receive care packages from family, but a gang of Ukrainian prisoners stole anything of value. The slaves took to sleeping with their gifts, and even their shoes, under their heads.

Von Braun kept his top staff at Peenemünde during the construction of Mittelwerk. That meant the German engineers supervising the underground work at the new site weren't always the most competent. Sellier described one confused German engineer whose job it was to supervise the laying of rail lines in the tunnels. The prisoners would lay tracks according to the supervisor's specifications, only for the German to change his mind and order the slaves to rip up the rails and shift them a few feet in one direction or another.

In February 1944, Himmler called von Braun to a meeting. There, Himmler not-so-subtly implied that von Braun was deliberately holding up the V-2's development. 'I trust you realize that your … rocket has ceased to be an engineer's toy,' Himmler said. Von Braun knew the truth – that the V-2 was late in development because it didn't yet work. Allied bombing, and the broader collapse of Germany's war effort, certainly weren't helping. Himmler may very well have known the truth, as well. It's possible he saw delays inside the V-2 effort as an opportunity to increase the SS's control of the program and decrease the army's own influence. 'Why don't you come to us?' Himmler asked.

Von Braun rebuffed the SS chief. He compared the V-2 to a flower that needs sunlight, fertile soil and a gardener's loving care. If the SS poured fertilizer on the rocket program in a desperate effort to make it grow fast, the paramilitary group just might kill it. Tensions rose between von Braun and his SS overlords. In March 1944, von Braun got drunk at a party with his brother Magnus and some associates. Germany was losing the war, the sloshed von Braun groused. The Reich was wasting von Braun's talents on all these vengeance weapons, he said.

Instead, he should be working on a 'spaceship', he said.

Someone snitched. SS police briefly detained von Braun, his brother and two other partygoers. Dornberger intervened and, in reminding top Reich officials of von Braun's importance to the V-2 program – indeed, von Braun was the V-2 program secured the men's release. There's not a lot of surviving documentation detailing the arrests and it's not even clear that von Braun's drunken 'defeatism' – to borrow the SS's term – was the real reason for the detentions.

Six months later on 8 September, 1944, special army battalions in The Netherlands and the Ardennes forest in Belgium finally launched the first operational V-2s. One arced toward Paris, which invading Allied troops had just liberated. Two more V-2s streaked toward London. Von Braun and his team celebrated with champagne.

The camouflaged entrance to the Mittelwerk underground rocket factory. The SS pulled Russian, Ukrainian, Polish, Czech, French and Italian prisoners from the Buchenwald concentration camp and other prisons both to build Mittelwerk and, later, to operate its underground plants. The slaves lived in a new camp called Mittelbau-Dora. All 16,000 of them. (US National Archives photo)

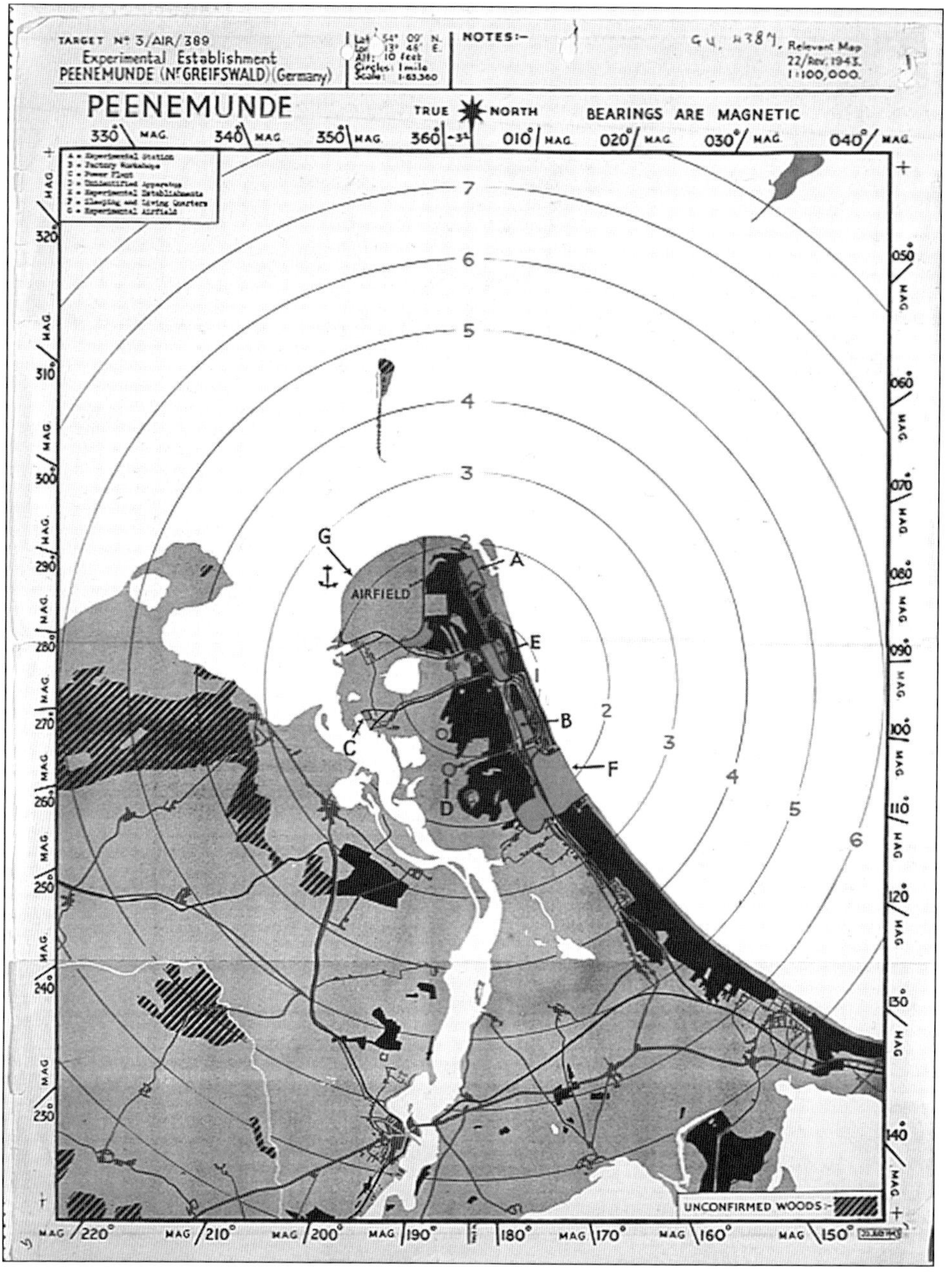

Six hundred Royal Air Force bombers struck the V-2 facility at Peenemünde on 17 and 18 August, 1943, badly damaging the site and setting back, by months, Wernher von Braun's progress toward a functional weapon. Hundreds of slaves died. This is the RAF's target map for the raid. (UK National Archives photo)

Between 8 September 1944 and 27 March 1945, no fewer than 3,172 V-2s struck Belgium, France, The Netherlands and the United Kingdom. In all, V-2 strikes killed around 9,000 people. (German Federal Archive photo)

Chapter 6

V-2s rain on Europe

That September 1944 launch was the culmination of decades of work by von Braun and his colleagues. It would end up costing the Third Reich $30 billion in 2021 dollars – and also cost thousands of prisoners their lives. As many as 20,000 workers perished while producing V-2s at a peak rate of around 20 per day.

Between 8 September 1944 and 27 March 1945, no fewer than 3,172 V-2s struck Belgium, France, The Netherlands and the United Kingdom. The Germans even launched a few at a bridge in Germany in a failed attempt to prevent the Allies from crossing the span.

A nine-year-old boy named Eric Taylor and his sisters Pat, 12, and Jill – just two years old – witnessed one of the first V-2 raids on London in the fall of 1944. 'We were well accustomed to air-raids and V-1 'doodle-bug' raids, as we lived in a prime target area,' Taylor told the BBC in 2003. 'However, there was no air-raid warning given out on that particular morning.'

The siblings arrived at Jill's daycare in the Dagenham Borough in East London before 7a.m., pushing their baby sister in a stroller. Normally, the toddlers would play on the playground outside the school which opened at seven, but for some reason on that particular morning, the staff brought the kids indoors early. Decades later, Taylor would call it a 'miracle'.

Having dropped off Jill, the older siblings pushed the empty stroller outside. 'Pat clearly remembers looking back and admiring the straight parallel tracks made in the dew on the grass by the wheels of the now-empty pushchair,' Taylor recalled. 'Then there was a tremendous explosion.'

Thick black smoke curled into the air. 'Slowly-swirling debris, which included a whole wooden door, rose straight up from the edge of the playground up into the clear sky.' People were running around, talking about sabotage. No one had heard of the V-2. No one had experienced a Nazi weapon that struck without warning – and without a sound. Taylor's mom was at work at a nearby lumber yard when the rocket struck. She heard the explosion and saw the smoke rising from the blast site near the school. It was 50 years before Taylor asked his then-elderly mother about that day.

It must have been horrible, Taylor said. 'It was,' his mother said. She recalled mothers racing along a lane toward what they surely

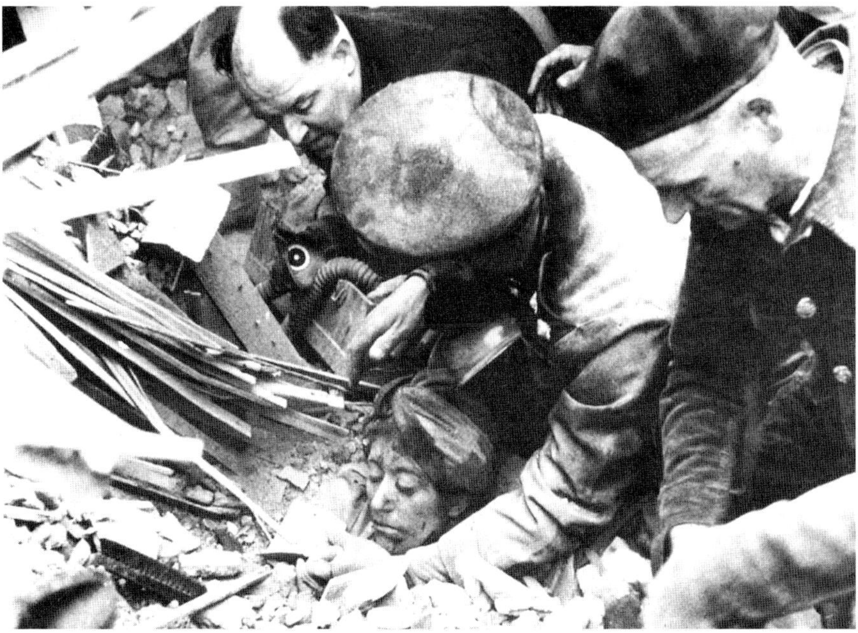

Firefighters from the London Fire Brigade rescue a woman buried up to her neck in debris following a V-2 strike on the London neighborhood of Manor Park. She survived, but around 9,000 other victims of V-2s were less fortunate. (London Fire Brigade photo)

imagined was a scene of terrible bloodshed. But incredibly, the V-2 had killed no one. Taylor's mom fetched Jill then went looking for Eric and Pat. In shock, the siblings had wandered away, down 'rubble-strewn side-streets', still pushing Jill's empty stroller. A passing ambulance crew eventually picked them up. 'The episode has had its effects on me,' Taylor wrote in 2003. 'I'm still afraid of sudden loud bangs. But, in a way, I don't mind. It reminds me, after all these years, of the time when a miracle spared the lives of so many innocents.'

There were other 'miracles'. In February 1945, a V-2 prematurely exploded over Grosvenor Square in London. A US Army major watched the debris rain down around him. His name was Robert Staver. In a dramatic twist worthy of a movie, the Army had assigned Staver to commandeer as many V-2s as he could and ship them to the United States for study, duplication and deployment.

A V-2 struck Hazelhurst Road in Tooting in South London on 19 November 1944. Thirty-five people died. There's no evidence Wernher von Braun appreciated the scale of the bloodshed his rockets inflicted – or, if he did, let it bother him. A colleague shrugged off von Braun's ambivalence. 'We were at war,' the man said, 'and although we weren't Nazis, we still had a Fatherland to fight for.' (London Fire Brigade photo)

Chapter 7

Dreams of outer space

The arc of the war in Europe was clear but Japan remained undefeated. The V-2 could be useful to US forces as they closed in on the Japanese Home Islands.

Privately, von Braun told his fellow scientists that the V-2 wasn't much of a weapon and couldn't alter the trajectory of the war. He was right, of course but in the process of failing, the V-2 killed around 9,000 people. On 16 December 1944, a V-2 slammed into Cinema Rex in Antwerp. A staggering 567 people died under the collapsing roof.

There's no evidence von Braun appreciated the scale of the bloodshed – or, if he did, let it bother him. A colleague shrugged off von Braun's ambivalence. 'We were at war,' the man said, 'and although we weren't Nazis, we still had a Fatherland to fight for.'

While his beloved rockets were inflicting death and destruction across Europe, von Braun increasingly fixated on his dream of exploring outer space. 'Let's not forget,' von Braun told his team, 'that this is only the beginning of a new era, the era of rocket-powered flight. It seems that this is another demonstration of the sad fact that so often important new developments get nowhere until they are first applied as weapons,' he added.

Even as V-2s rained down on Allied cities, von Braun found ways to sneak purely scientific studies into the rocket program. In early 1945, he provided rockets to a team, led by physicist Erich Regener, that was working on high-altitude meteorological instruments. Regener never actually got to launch his meteorological V-2. Russian advances toward the Baltic sent the physicist and his team fleeing deeper into Germany.

The writing was on the wall. On October 1, 1944, the American First Army captured Aachen, the first German city to fall to the Western Allies. German troops fell back. On 20 October, Soviet troops and Yugoslav partisans liberated Belgrade. A month later, the Soviets had enveloped Budapest. The Germans fell back.

German defenses stiffened along the borders of Germany proper and in mid-December the Reich's forces staged a bold, but ultimately doomed, counter-offensive in the Ardennes. But more and more Allied troops poured into Europe in the west and Soviet strength swelled in the east. All the while, eight V-2s struck London every day, on average – killing hundreds and, just as von Braun had warned, doing absolutely nothing to save the Reich from defeat.

As Soviet troops closed in on Peenemünde in the first weeks of 1944, the SS ordered thousands of workers to take up rifles and prepare to defend the facility. That final battle never came for the Peenemünde staff. They evacuated instead. Meanwhile, the rocket battalions were losing their staging grounds to Allied troops. V-2 launches fell off.

At the urging of Hans Kammler, the SS's top weapons-developer, von Braun dusted off blueprints for a new version of the V-2 with large wings that allowed it to glide on re-entry and thus extend its range to as far as 340 miles. Nazi leaders hoped the winged V-2 would allow German forces to continue bombarding England even as they retreated deeper into Germany. It truly was a weapon of pure, desperate, pointless vengeance. Von Braun gazed on the winged V-2 and saw something else, entirely – the basis for the first manned spacecraft. Add a cockpit and one very brave pilot, and you could boost a human being just outside the atmosphere and still give him a chance of landing safely.

By late October 1944, von Braun had finished five winged V-2s. The first launch, on 27 December, flopped. The rocket's steering failed shortly after blast-off and it fell to the ground. The second test was more successful. The rocket-glider climbed to an altitude of 50 miles, tipped over and began to glide, then a wing broke off and the rocket tumbled to its destruction. But the brief glide surely gave von Braun hope that he might be able to develop his winged V-2 into a manned spacecraft.

However he had to defer that particular dream. Supply chains were collapsing as Allied armies overran mines, refineries, factories, railroads and highways. The supply of ethanol fuel was particularly dire for the German rocketeers. Further V-2 testing was impossible after the New Year. The front-line rocket battalions managed just a few more operational launches of V-2s before the SS ordered them to the front to die as infantry.

A V-2 engine in the underground factory at Dora-Mittelbau in May 1945. While his beloved rockets were inflicting death and destruction across Europe, Wernher von Braun increasingly fixated on his dream of exploring outer space. 'Let's not forget,' von Braun told his team, 'that this is only the beginning of a new era, the era of rocket-powered flight.' (US National Archives photo)

By late October 1944, Wernher von Braun had finished five winged V-2s. Nazi leaders hoped the winged V-2 with its 340-mile range would allow German forces to continue bombarding the United Kingdom even as they retreated deeper into Germany. (Smithsonian National Air & Space Museum photo)

In late January 1945, as Soviet tanks were rolling westward across Poland, von Braun called a secret meeting of his closest colleagues at a farmhouse. 'Germany has lost the war,' he said, according to a ghostwritten first-person essay from 1958. 'But let us not forget that it was our team that first successfully reached outer space. We have never stopped believing in interplanetary travel. Each of the conquering powers will want our knowledge,' von Braun said. 'The question we must answer is: to what country shall we entrust our heritage?'

The answer was obvious to von Braun. 'Our decision was unanimous,' he wrote. 'We would surrender, it was agreed, to the American army.'

Von Braun 'was an opportunist, of course,' historian Roger Launius said, 'but he was hardly willing to work for anyone.'

Chapter 8

Mittelbau-Dora

For von Braun, surrendering with his team intact was easier said than done, of course. As Allied troops closed in, the SS directed von Braun and the other V-2 managers to speedily relocate their organizations to the Bavarian Alps, a rugged redoubt that Nazi diehards could hope to defend for many months to come. At the same time, there were rumors that the SS might shoot the V-2 managers once the Allies were upon them.

Before departing south, von Braun paid one last visit to the prisoner camp at Mittelbau-Dora. It was overcrowded with starving prisoners the Nazis had transferred from camps that lay closer to Allied lines. So many prisoners died at Mittelbau-Dora in the war's final months that camp administrators gave up trying to cremate their bodies in orderly fashion and simply tossed them on open pyres.

In March, the long-time Ukrainian and Russian residents of the camp rose up. Guards put down the rebellion and hanged the rebels. Two mass-hangings even took place inside the Mittelwerk tunnels, on a crane the workers had used to lift V-2s.

Von Braun was in a car accident around this time. His arm was badly broken and his mobility was limited. There always seemed to be an SS guard at his elbow and the scientist grew paranoid. Whenever he heard Allied bombs falling and exploding, he feared the SS might decide the end was nigh – and send someone to kill him. In fact the assassin never came but an army messenger did. He summoned von Braun to a luxury hotel on the German-Austrian border where

Dornberger had taken up residence. There, von Braun enjoyed the fine dining and extensive wine collection … and waited.

On April 11, US Army scouts under the command of 1st Lieutenant Al Ungerleider crept up on the SS camp at Dora-Mittelbau.

'As we neared the camp, we started to get machine gun fire from two small towers in the distance,' recalled Captain Edward Burke, an officer in a tank-destroyer battalion supporting Ungerleider's scouts. 'Two of my tank-destroyers destroyed the towers and my tank-destroyer along with another crashed through the front gates,' Burke said. 'The machine-gunners were killed, we captured about 44 prisoners and the rest of the Germans escaped.'

The Americans entered the camp and what Ungerleider saw 'burned into my brain and my soul like nothing else in my life,' he said later. 'My men and I smashed through the gates and witnessed the site of dead bodies, of human beings in the worst state of degradation,' Ungerleider said. 'There was absolute horror in what we saw.'

At its peak capacity, Dora-Mittelbau had housed 16,000 people. By mid-1945 the Germans marched out all but a few hundred starving slaves – but they also left behind a thousand bodies they didn't have time to burn.

'The sight was absolutely horrifying and the smell was unbearable,' Burke said. 'The prisoners looked like walking cadavers.' Ungerleider and one of his soldiers surprised a guard in one building and shot him dead before he could open fire.

'Back outside there was also a railroad siding with two empty boxcars,' Burke said. 'Their cargo of dead bodies had been stacked up like cordwood, four to five feet high against the back wall of the crematorium. The smell was just awful.'

Furious, Ungerleider marched into nearby towns and ordered, on threat of imprisonment, the German residents to take food to the camp. But it was too late for many of the prisoners, who died the same day they were liberated. Ungerleider was Jewish and spoke

Yiddish. He gathered up some of the stronger prisoners and together they recited the Kaddish, the Jewish prayer of mourning.

With the Allies advancing toward Mittelbau-Dora in early April, the SS had begun pulling prisoners from the camp and marching them deeper into German territory. As one procession of thousands of Russian, Polish, French, Dutch, Hungarian, Italian prisoners reached Gardelegen, 90 miles west of Berlin, the SS and army guards grew impatient. Allegedly following orders from local Nazi official Gerhard Thiele, on April 13, 1945 the guards packed 1,016 of the weakest prisoners into a barn … and set it on fire.

The men died in heaps. They died trying to claw through the narrow gap between the barn wall and the ground. They died begging for help in a half-dozen languages.

The next day, a US Army patrol happened across the scorched barn. 'I was about 15 miles away,' soldier Henry Skwirut recalled decades later. 'My sergeant, Sam Coscia, and I jumped in my Jeep and drove over. I'll never forget it,' Skwirut said. 'The smell of burning flesh was horrible, and the bodies were everywhere. We weren't fearful, because at that time, we knew the Germans were on the run. It was all almost over. But I felt angry. Angry for what they had done.'

Hitler shot himself in his Berlin bunker as Russian artillery demolished the city on April 30. As the Third Reich collapsed, von Braun sat in his hotel, waiting and wondering. The French army was a few miles to the west. The Americans were a few miles to the south. But no one came.

At the same time, the US Army – as part of its Special Mission V-2 – dispatched four teams, including Staver's, to track down all the V-2s and parts of V-2s that they could find. The teams had their work cut out for them. The British and Soviets had their own V-2-recovery squads and in these heady days of collapse and occupation, there weren't a lot of rules. It was finders, keepers.

Staver had his eyes on Mittelwerk. He arrived at the US-occupied facility on May 12. He'd heard a Soviet team was en route and would arrive in a little over two weeks. The Soviets would be in their rights to kick out the Americans. As the Allies carved up Germany into occupation zones, Moscow had leaned on Washington and London for control of the area around Mittelwerk. Two weeks, that was Staver's window – and he made good use of it. By the deadline, he'd loaded 400 tons of rocket parts onto train cars, enough for a hundred V-2s.

The Americans had a lot of parts, but no idea how to assemble them into a functioning V-2. After loading up his last railcar, Staver drove all around Mittelwerk in a jeep laden with cigarettes, alcohol and cans of preserved meat – perfect bribes in post-collapse Germany. A series of sources led Staver to Karl Otto Fleischer, one of von Braun's scientists. Fleischer eventually pointed Staver to a huge stash of V-2 documentation – 14 tons of it – in a mine near Mittelwerk.

Time was short. Staver needed 10-ton trucks to haul the crates full of documents from the mine, but the approving authorities were in Paris. So Staver cajoled a P-47 pilot to let him squeeze into the space behind his fighter's single seat. They sped to Paris at treetop height and it was there that Staver made a fateful decision – one that would set in motion Operation Paperclip, the US government's crash effort to recruit German rocket-scientists.

The operation would have epoch-making implications. The ex-Nazis would be responsible for creating, or *recreating*, the technology that would carry American explorers into space. The same tech would extend the reach of America's atomic arsenal over every square mile of the planet. Staver told his colonel in Paris that he'd secured V-2 parts and was about to secure the instructions for assembling them, but there was a third part to the Germans' rocket program, he explained. The scientists. Staver asked for permission to recruit ex-Nazi rocketeers. 'You write the cable and I'll sign it,' the colonel said.

The administration of President Harry Truman soon formalized Staver's effort and gave it a name: Operation Paperclip. The White House however banned the program from recruiting active members of the Nazi Party, but the officials running Paperclip knew that would render the program open to debate. So they just shrugged ... and lied.

Above: At peak capacity, the Dora-Mittelbau prison camp housed 16,000 people. By mid-1945 the Germans had marched out all but a few hundred starving slaves. But they also left behind a thousand bodies they didn't have time to burn. In this photo from April 1945, a Polish survivor shows US Army soldier John Lyndon the crematorium ovens the SS used to burn corpses at Dora-Mittelbau. (John Driza photo)

Opposite above: After liberating the Dora-Mittelbau prison camp in April 1945, the US Army forced the German residents of nearby Nordhausen to carry to mass burial sites the bodies of prisoners who died in the camp. (United States Holocaust Memorial Museum photo courtesy of Joseph Mendelsohn)

Opposite below: In April 1945, US Army troops supervise German civilians burying the bodies of hundreds of prisoners who died in the Dora-Mittelbau prison camp. 'The sight was absolutely horrifying and the smell was unbearable,' US Army officer Edward Burke said. (US National Archives photo)

A US soldier looks into a crematorium at a Nazi concentration camp in May 1945. Human remains are visible in the oven. So many prisoners died at Mittelbau-Dora in the war's final months that camp administrators gave up trying to cremate their bodies in orderly fashion and simply tossed them on open pyres. (US National Archives photo)

With the Allies advancing toward Mittelbau-Dora in early April, the SS had begun pulling prisoners from Mittelbau-Dora and marching them deeper into German territory. As one procession of thousands of Russian, Polish, French, Dutch, Hungarian, Italian prisoners reached Gardelegen, 90 miles west of Berlin, the SS and army guards grew impatient. Allegedly following orders from a local Nazi official, the guards packed 1,016 of the weakest prisoners into a barn … and set it on fire. (Charles and Katie Overstreet photo courtesy of the Flora Public Library)

Chapter 9

The rocket scientists meet the Americans

By May 2, 1945, Dornberger and von Braun were tired of waiting. They dispatched von Braun's brother Magnus, who spoke good English, on a bicycle to go find the Americans. Magnus ran into a US Army anti-tank platoon and spent half an hour convincing them that a bunch of the Third Reich's top rocket scientists were waiting in a nearby hotel. An intelligence officer belatedly handed Magnus some passes for safe passage and told him to produce the scientists. Von Braun and Dornberger selected four other men to accompany them and Magnus. They piled into staff cars and threaded mountain roads until they encountered an American checkpoint.

The GIs inspected the passes and, deciding the Germans were who they said they were, escorted them to the town of Reutte, where the US Army's Counterintelligence Corps lodged in a mansion. Von Braun later said he half-expected the Americans to kick him in the teeth. Instead, they scrambled up some eggs, boiled some coffee, put out bread and butter and served the Germans breakfast.

Von Braun and his associates were already on a list of high-ranking German officials the Americans wanted to talk to. They stayed overnight, enjoyed another luxurious breakfast and posed for a lot of photographs with curious, amused American soldiers. Von Braun even spoke to a US Army reporter. In the interview, he insisted that, with another two years of work, he could have improved the V-2's guidance system to 'pinpoint accuracy' and boosted Mittelwerk's production rate to 200 rockets a day. It would have been enough to alter the course of the war, he claimed.

The next day, the Americans drove von Braun and the others to Garmisch-Partenkirchen, a ski resort that had hosted the 1936 Olympics. Now the resort was hosting a reunion of sorts as more and more German rocket scientists arrived with their American handlers. Soon the US Army had assembled roughly half of von Braun's science and engineering team from Peenemünde.

An alphabet soup of US intelligence agencies took turns interviewing the scientists. There would be many, many more debriefings in following months, both in Europe and in the United States. The Germans spilled their guts. Von Braun for his part spoke at length about two related ideas. A space-plane – in essence, a more powerful version of the manned V-2 he had tinkered with a few months earlier – and an orbital 'observation platform'. A space-plane – a 'multi-stage piloted rocket' in von Braun's phraseology – could accelerate to a top speed of 24,600 feet per second, or 4.7 miles per second, von Braun told his interrogators, according to a 1951 paper the scientist wrote based on his recollections of the conversations.

'At such speeds the rocket would not return to Earth, as gravity and centrifugal force would balance each other out,' von Braun wrote in the paper. 'The rocket would complete one circuit around the Earth in any time between 1.5 hours and several days. The whole of Earth's surface could be continuously observed from such a rocket. The crew could be equipped with very powerful telescopes, and be able to observe even small objects, such as ships, icebergs, troop movements, construction work, etc.'

'The importance of such an "observation platform" in the scientific, economic and military sphere is obvious,' von Braun continued. The crew of the observation platform could return to Earth by simply decelerating and letting gravity pull them down to the surface. 'It can land like an ordinary airplane by means of wings and auxiliary gear.'

For an around-the-clock presence in orbit, swap the space-plane for a permanent orbital station, von Braun advised. 'The erection could be easy, as the components would have no weight in the state of free gravitation. The work could be done by men who would float in space, using divers' suits, and who could move at will in space by means of small rocket propulsion units, the nozzles of which they would point in the required direction.'

These twin ideas – space-plane and space station – might have sounded like fantasy to the agents interrogating von Braun. In fact, both ideas inspired formal military programs in the late 1950s and early 1960s. Those programs failed, but they led directly to separate, civilian efforts that eventually produced both working space-planes – in the form of the Space Shuttle – and functional space stations, including Skylab and the International Space Station.

It's inaccurate to say these vehicles lacked any military application. The Space Shuttle in particular performed important, albeit largely secret, military roles but they weren't weapons. Which is not to say the largely peaceful applications of von Braun's ideas were inevitable. For in the same conversation where he accurately sketched the next 50 years of space exploration, von Braun also described how humanity might weaponize space. He mentioned Oberth and Oberth's own fictional station. 'An observation station of this type could be equipped with an enormous mirror, consisting of a huge net of steel wire onto which thin metal foils could be suspended,' von Braun wrote.

'A mirror of this nature would have a diameter of many kilometers, and its component facets could be controlled by the station, which would enable the heat and light of the sun to be concentrated on selected points of the Earth's surface. This would enable large towns, for instance, to get sunlight during the evening hours. The weather, too, can be influenced by systematic concentration of the sun's rays onto certain regions. Rain could be induced to fall on regions hit by drought, by concentrating the sun's rays onto distant lakes and seas, and increasing their evaporation. The clouds thus formed could be driven to the required spot by influencing the centers of low and high pressure.'

Von Braun went on about orbital weather-control for a while before pivoting to the obvious implication of his, and Oberth's, conjectural giant space mirror. Oberth had only hinted at his space station's city-destroying potential. Von Braun dispensed with the fig-leaf of the station's possible peaceful applications. 'If the mirror is made large enough, and it could be of extraordinarily light construction, it would even appear possible to generate deadly degrees of heat at certain spots of the Earth's surface.'

On August 6 and 9, 1945 the United States dropped atomic bombs on two cities in Japan, hastening the end of the Pacific War. Von Braun was aware of the scientific and military implications of the destruction of Hiroshima and Nagasaki – indeed, he idly described to his Army interrogators a possible nuclear-powered spacecraft. However he did not propose arming his space-plane or space station with nuclear weapons. *At least not yet.*

There was quite a lot of talk of orbital weapons of mass destruction as US Army agents interviewed former Third Reich rocket scientists in Germany in mid-1945. It's possible von Braun wasn't the only Peenemünder waxing lyrical about frying cities from space but someone – it's not clear who – outlined more detailed plans for the giant space mirror that von Braun described in his own recollections of the post-war debriefing. The mirror in this detailed scheme would be 3.5 square miles in area, made of metallic sodium and orbiting 5,000 miles over Earth's surface. The mirror would attach to pressurized living quarters, supplied via regular visits from manned rockets. Thousands of pumpkin plants, each inhaling carbon monoxide and exhaling oxygen, would provide the atmosphere.

Life published illustrations of this so-called 'sun gun' in its July 23, 1945 issue. 'The Germans, the Army reported, hoped to use such a mirror to burn an enemy city to ashes or boil part of an ocean,' *Life* claimed.

Above: The Allies tried 19 ex-Nazis for war crimes they committed at Dora-Mittelbau. Sixteen of the 19 defendants are depicted in the September 1947 photo. The Allies notably did not try Warner von Braun or any of the other Peenemünde rocket scientists. (US National Archives photo)

Opposite above: Wernher von Braun worked hard to obscure his Nazi past and the US government helped. To sell von Braun as a wholesome embodiment of middle-American values, the US military and NASA took pains to put von Braun in front of children, regardless of how stiff and uncomfortable the scientists looked in public settings – and especially around kids. (NASA photo)

Opposite below: The concept of a giant space mirror – one with obvious destructive potential – dates as far back as Hermann Oberth's 1923 book *The Rocket into Planetary Space.* The idea reappeared in 1945 as Allied agents interrogated ex-Nazi scientists about the Third Reich's rocket program. Someone – it's not clear who – outlined detailed plans for a giant space mirror. The mirror in this scheme would have been 3.5 square miles in area and made of metallic sodium. Orbiting 5,000 miles over Earth's surface, it would have illuminated, warmed or fried entire cities. The space-mirror concept endured. As late as the 1970s, companies pitched orbital reflectors to NASA as a method of climate-control. (NASA art)

The article, as well as a similar piece that appeared in *Time* that summer, is incredulous, sensationalistic and even a little silly. The Third Reich couldn't have built a 3.5-mile-square space mirror even if it had devoted all of its resources to the undertaking for decades. But at the same time, the *Life* and *Time* stories accurately captured a sentiment among the ex-Nazi rocket scientists. That mankind could – and would – escape Earth's gravity, establish a presence in space and, there in the cold and airless dark, assemble weapons of unimaginable power.

If anything, the Germans were too gentle in their conceptions of orbital warfare. In fact, the weapons they would help to create in coming decades would greatly exceed the destructive potential of some unwieldy space mirror.

Today the world's major powers possess tens of thousands of nuclear warheads, each capable of wiping out a town or city faster, and at less cost, than any sun gun could do. Those warheads travel through space on rockets that von Braun and his colleagues invented, almost as an afterthought, while realizing the space-planes, space stations and other artifacts of the science fiction they'd read as kids.

PART TWO

THE DESERT

Old rockets,
some orbiting balls of metal
and a bunch of tragic space dogs

Chapter 10

Across the Atlantic

In the fall of 1945, British intelligence asked to 'borrow' von Braun, Dornberger, Rudolph and a few other V-2 scientists before the Americans whisked them away across the Atlantic. The British put the Germans to work prepping and launching four V-2s at a target in the North Sea. London hoped to glean from the launches the key technical data behind the weapon that had so cruelly terrorized the British population in the war's final months.

It seems some of the British technicians on the V-2 tests were willing to forgive and forget. They and the Germans tapped the project's supply of ethanol and got roaring drunk together. A British officer caught the group stumbling around the test site, arm in arm, singing the German war anthem 'We Will March Against England'.

Dornberger, whose use of slave labor was well-known, wasn't invited to the party. The British kept him in a separate lodging under armed guard. When the V-2 tests ended, British agents shipped von Braun and Dornberger to London for debriefing. The Americans showed up a few weeks later to fetch their prized Germans. The Brits let von Braun go but they declared Dornberger a prisoner of war and held onto him. 'The British pulled a sneaky on us,' Staver said.

Dornberger wound up in a prisoner-of-war camp in Wales. He was an unpopular prisoner, even among his fellow Germans. He had a slaver's stink on him. During his imprisonment, Dornberger related to British interrogators a story that von Braun's leading biographers have failed to confirm. Dornberger told his captors that, in the days following the V-2 scientists' surrender to the Americans – during a layover in the town of Witzenhausen, inside the American occupation zone – a team of Soviet agents attempted to kidnap von Braun. The Soviets showed up in Witzenhausen wearing British uniforms and carrying authentic security passes they had somehow obtained. The Americans smelled deception and sent the Soviets packing, Dornberger said. 'That's how those Russians operated, real kidnapping. They had no scruples at all.' Dornberger would spend two years in the prisoner-of-war camp before the Americans managed to pry him loose.

After retrieving von Braun from the British, the Americans split up the V-2 team. The top few members, including von Braun himself, flew to the United States – a three-day trip requiring several refueling stops and plane-changes. The rest endured a long journey by sea.

Apparently between stamping his passport in Delaware on September 20, 1945 and shipping off to the US Army rocket facility in New Mexico on October 8, von Braun spent some time at PO Box 1142, the same secretive Virginia facility where, two years before, interrogators had overheard German prisoners talking about Peenemünde. He wasn't a prisoner, but his knowledge of German rocketry was of acute interest to the intelligence agencies whose agents haunted PO Box 1142. Between conversations, von Braun's spooky hosts took pains to entertain him and his fellow scientists.

'I was the morale officer of Wernher von Braun,' Arno Mayer revealed in 2008, shortly after the government declassified much of the surviving documentation from PO Box 1142's wartime work. Mayer plied von Braun and the other rocket scientists with magazines and liquor. He also escorted von Braun and three colleagues into Washington, DC one day for a little Christmas shopping. The Germans said they wanted to buy underwear for their wives, who were still in Europe. So Mayer took the men to a Jewish-owned department store. 'We told the salesperson what size and so on,' Mayer recalled in an interview with National Public Radio. 'And the woman held up a pair of panties.' The Germans were 'appalled', according to NPR reporter Pam Fessler. They didn't want panties. They wanted long woolen thermal underwear, the kind that keeps you warm in the winter.

In late 1945, Wernher von Braun spent some time at PO Box 1142, the same secretive Virginia facility where, in 1943, interrogators had overheard German prisoners talking about the Peenemünde rocket facility – intel that may have led Royal Air Force bombers to the site. PO Box 1142 was located inside Fort Hunt, a mostly defunct base on property outside Washington, DC that once belonged to George Washington, America's first president. (US National Park Service photo)

Chapter 11

Captured V-2s

Around half of the 127 German rocketeers the Americans initially recruited under Operation Paperclip wound up at Fort Bliss in Texas, from where they frequently traveled by bus to White Sands Proving Ground in New Mexico for test-launches of captured V-2s.

The US Army's V-2 recovery program was a mess. Thanks to Staver and his jeepful of bribeworthy goodies, American forces had scooped up a not insignificant amount of V-2 components. But the Germans had not designed their rocket parts for longevity. In general, the launch battalions fired off a V-2 within a couple weeks of it leaving Mittelwerk's tunnels.

The parts the Americans shipped to White Sands were months old and rusted-over. The Army had contracted with General Electric to oversee the V-2 effort. It took company engineers eight months – working in rudimentary facilities that were sweltering hot during the day, cold at night and perpetually short on fresh water – to recondition and assemble the parts.

In what by now clearly was a trend for the crude rocket, the American V-2s didn't work very well. The first V-2 von Braun launched at White Sands, in April 1946, spiraled out of control. Later launches were more successful, however, and by the fall of that year, the Americans had begun packing simple scientific instruments in the noses of the reconditioned V-2s. In October, a V-2 carried a US Navy spectrograph to an altitude of 107 miles, high enough for the device to capture – and beam back to the ground – ultraviolet images of the sun. It was America's first foray into space.

Through 1946 and 1947, a few more of von Braun's German colleagues arrived in the United States and slowly congregated at Fort Bliss. But post-war demobilization was taking a bite out of the Army's rocket program. In 1945, the Allies had, with great urgency, tracked down, interrogated and recruited the Third Reich's rocketeers. Just two years later, the urgency left US military rocket programs like air escaping a punctured balloon.

In 1947, the administration of President Harry Truman, together with Republicans in Congress, gave the Pentagon a choice. Lawmakers would pay for new long-range bombers and high-tech new fighters *or* rockets and other experimental weaponry – but not both.

The military chose the bombers and fighters. Work on rockets slowed. A few over-eager soldiers at Fort Bliss lobbied von Braun to put them inside a V-2 and launch them into space but Von Braun said no. It's not that he didn't want to put a man in orbit. He just knew it couldn't be done here, like this, right now.

Von Braun found himself with less and less to do as G.E. gradually took over the V-2 program. A highlight of 1947, however, was his test-launch of a V-2 from a Navy aircraft carrier.

Von Braun still lacked official immigration status – to say nothing of a path to US citizenship – but in late November 1946 the War Department decided to lift the veil of secrecy around the Paperclip scientists. The Army invited reporters to Fort Bliss to interview von Braun and his associates. In the following weeks, von Braun appeared in photos in *The New York Times*, *The Washington Post*, *Time* and, of course, *Life* – purveyor of the 1945 Nazi space-mirror scare.

Around the same time, von Braun convinced the Army to bring his family over to the United States. He clearly intended to stay and later would describe the United States as 'the homeland of my free choice'.

Not coincidentally, in 1946 he was 'born again' at a tiny evangelical church in Texas. From this point on, von Braun's apocalyptic religious faith began to thread around his stiffening American patriotism, which itself wound around his absolute, lifelong belief in technology – specifically, rockets – as the vehicle of mankind's destiny. Or, at least, *his* destiny.

But von Braun realized that, if he was going to have a future in America as a creator of rockets and a pioneer of space, he needed strong government support and that meant selling his vision of space to millions of everyday Americans.

Decades removed from the Cold War space race, the formation of NASA, the Apollo Moon landings, the Space Shuttle missions, the construction of the International Space Station and the successful launch of far-flying probes to all corners of the solar system, it's easy to forget that, as recently as the late 1940s, Americans didn't give a damn about space. That attitude would change in profound ways within a generation and even today space exploration is widely popular. NASA might not enjoy the high levels of funding it had in the late 1960s, but at the same time *shutting down* NASA is inconceivable to most Americans. They accept, as a matter of course, that a strong and prosperous America is an America that not only explores space, it *leads* other countries in exploring space.

That assumption didn't come from nowhere. It has roots in the military impulse to occupy the orbital high ground that gained strength in the early years of the Cold War, but it's also a product of a cultural fixation on space.

Space has long been the subject of popular fiction. That popularity spiked in the 1950s, arguably inspiring growing government support for civil and military space programs. Von Braun was a key figure in that popularization of space. In other words, not only did Nazi Germany's top rocketeer seed the US space program with key technology, he also seeded the American imagination with ideas that inspired an enduring national effort to explore – and weaponize – space.

The US space program shares its technological DNA with the Nazi rocket program. By the same token, America's obsession with space began with, and to a great extent has been defined by, the ideas of a man who, at best, took advantage of Nazism. At worst, he himself was an actual Nazi.

Wernher von Braun and his German colleagues pose for a photo at Fort Bliss, Texas in March 1946. The Germans frequently traveled by bus to White Sands Proving Ground in New Mexico to test-launch V-2s that General Electric rebuilt from captured components. (US Army photo)

Chapter 12

A novel fails to launch

Right-wing racial and sexual chauvinism were the scaffolding propping up von Braun's turgid entrée into popular fiction. In 1947, during the lull in Army V-2 development, von Braun set out to write a novel about space. He wasn't the only one working on a book. His old mentor Ley, who had fled Germany for the United States back in 1935, had noted a conspicuous lack of enthusiasm for space exploration among everyday Americans. Ley set out to change attitudes. His 1949 book *The Conquest of Space* signaled a sea change. His book included detailed descriptions of space stations and manned flights to the Moon.

Von Braun finished his own book, *Mars Project*, in 1949. It was … not good. He wrote it in German, some of his German colleagues helped with the science in the book and a friend of his mother translated it into English. American publishers were befuddled. A German publishing house circulated a heavily edited version of the book in the 1950s, but it wasn't until 2006 that a Canadian trade publisher finally released the original novel in English.

The story illuminates von Braun's technological and ideological interests. In his version of 1980, a US-led alliance has finally defeated the Soviet bloc. The key to victory: an Earth-orbiting space station which von Braun named 'Lunetta' just like in his 1930 short story that rained atomic bombs on enemy cities. In the aftermath of its atomic victory, the American bloc points a huge space telescope at Mars and discovers evidence of civilization.

The Americans and their allies launch gigantic, reusable rockets, each twice as big as the record-setting Saturn V rocket that von Braun eventually would develop and which would be his most enduring technological legacy. In nearly a thousand round-trips, the rockets deliver into orbit the components of a fleet of ten interplanetary spaceships. The 70 crew – all male, as von Braun was irredeemably sexist – lands on Mars aboard gliders. That plot point assumes, as many people did in the mid-twentieth century, that Mars still clings to a dense, Earth-like atmosphere.

Traversing Mars in wheeled rovers, the explorers discover an underground civilization of human-like beings. Their similarity to *homo sapiens*, von Braun explains, reflects God's design for the universe. Oddly, the communists in von Braun's tale don't warrant the same consideration.

The novel wasn't the flagship of space outreach von Braun had hoped it would be, but the ideas in the manuscript fared better than the manuscript itself did. Between 1951 and 1954 *Collier's* – one of the most popular periodicals in America, with a circulation of around four million – highlighted von Braun's vision in a series of lavishly-illustrated stories about space travel. Von Braun's fame began with the fawning new coverage in 1946 and swelled with the launch of the *Collier's* stories, but it was a series of TV films that von Braun made with The Walt Disney Company between 1955 and 1957 that made him a household name.

Disney producer Ward Kimball discovered von Braun in the pages of *Collier's*. He called up the scientist with a job offer and Von Braun didn't hesitate. He advised on, and appeared in, three Disney films: *Man in Space* and *Man and the Moon* in 1955 and *Mars and Beyond* in 1957. The movies explained the physics of spaceflight, sketched the basic design of a space station and explained how future explorers could travel to Mars aboard an atomic-powered spaceship. They were wildly popular and they studiously avoided any mention of von Braun's association with Nazis. Von Braun's rehabilitation – from a calculating Nazi rocket-maker to loveable American space hero – was well underway.

But to millions of Americans who were just learning about the patrician scientist with the thick German accent, it wasn't really a rehabilitation at all for they never knew von Braun had a dark past he needed to escape. In interviews and ghost-written, first-person magazine articles through the early 1950s, von Braun either omitted his SS and Nazi associations or fudged the details to cast himself as a victim of the Reich. In his own retelling, he was but a starry-eyed young man who fell under the Nazis' sway but who, in his heart, only ever desired to do good for freedom-loving people.

The US government assisted with this projection by publishing whitewashed official biographies of von Braun, but von Braun himself did most of the work of creating a new persona. 'We wouldn't have treated your atomic scientists as war criminals,' von Braun observed, 'and I didn't expect to be treated as one.'

The ex-Nazi's self-hagiography reached absurd heights in 'Why I Chose America', an essay von Braun and a ghostwriter penned for a 1952 issue of *The American Magazine*, a periodical for women. As the war ended, von Braun wrote, 'I faced the problem of choosing in my mind a custodian for the machines and the knowledge we had developed at Peenemünde.'

Should he go to America or the Soviet Union? Should he choose democracy or autocracy? 'Whereas democracies ask how can we reconcile our machines with basic concepts of individual freedom, the Russians are indifferent to human freedom,' von Braun wrote.

'Their rulers are using the power for machines to impose a police state on as much of Asia and Eastern Europe as they can,' he continued. 'This difference is why my choice was the West. The reason I chose America in particular is that Americans had a reputation for having an especially intense devotion to individual freedom and human rights. Furthermore, their traditional disinterest in conquest and their careful system of checks and balances in government offered the highest guarantee that any knowledge we entrusted to them would not be used wantonly.'

The fictional von Braun that lived in magazine articles and Disney movies – and thus in the American imagination – didn't willingly associate with the SS, the Nazi Party and indeed with Hitler himself in order to advance his rocketry. He didn't build weapons that wantonly killed thousands of people for no military benefit. He didn't drink champagne to celebrate the first V-2s that slammed into Allied cities. He didn't enjoy the free labor of tens of thousands of slaves. He didn't do all of this in pursuit of some grandiose vision of space exploration he had taken from a book he'd read as a teenager.

Von Braun's obsessions – liquid-fuel rockets, manned spacecraft, space stations, orbital weapons – never changed. But their context did. When von Braun was a Nazi, or at least a Nazi *associate*, his technological ambitions advanced the epoch-making aims of an unabashedly genocidal master race. After he became a God-fearing, Communist-hating American in the employ of the US Army, the very same tech served different ends. They were weapons for freedom, liberty, capitalism, evangelism and the unspoken white supremacy that girded American civic life. The rockets, space stations and sun guns didn't care who paid for them, built them and made use of them. Nor, really, did their creator.

In the early months of his rehabilitation, von Braun slipped up. In 1950, he wrote a memoir for the British Interplanetary Society. It was never published in its original form, which was good for von Braun's image-making, for in the manuscript, von Braun states plainly what should be obvious from an honest account of his work.

'We felt no moral scruples about the possible future use of our brainchild,' he wrote, describing his rocket group's decision to work for the German army. 'We were interested solely in exploring outer space. It was simply a question with us of how the golden cow could be milked most successfully.'

When the British Interplanetary Society finally published an abridged version of the memoir in its journal, those damning sentences had somehow disappeared.

A series of TV films that Wernher von Braun made with The Walt Disney Company between 1955 and 1957 made him a household name. Disney producer Ward Kimball had discovered von Braun in the articles the scientist wrote for *Collier's* and called up with a job offer. Von Braun didn't hesitate, he not only craved fame, he understood that his fame would benefit the US space program and, by extension, advance his own space ambitions. Pictured here, Walt Disney and von Braun at Redstone Arsenal in 1954. (NASA photo)

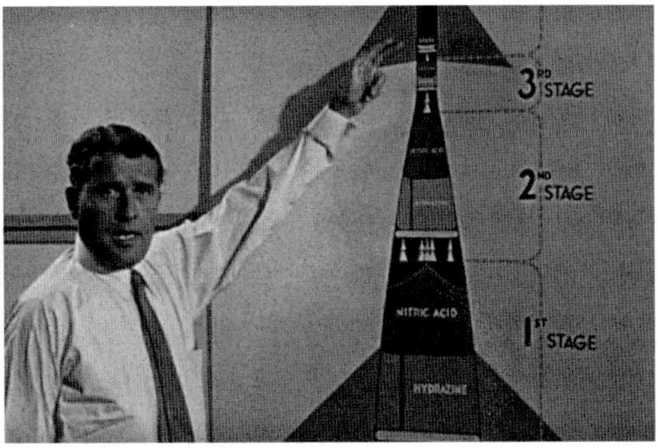

Between 1955 and 1957, Wernher von Braun advised on, and appeared in, three Disney films: *Man in Space*, *Mars and Beyond* and *Man and the Moon*, the last depicted here. The movies explained the physics of spaceflight, sketched the basic design of a space station and explained how future explorers could travel to Mars aboard an atomic-powered spaceship. (The Walt Disney Company capture)

Chapter 13

Redstone and Jupiter

The pause in US military rocket-development ended on August 29, 1949, the day the Soviet Union tested its first atomic warhead. Truman ordered the Pentagon to prioritize development of a new hydrogen bomb – a lighter and more powerful kind of nuke that could, in theory, fit in the nose cone of a rocket with intercontinental range.

The US Army and US Navy launched several overlapping and competing programs to develop increasingly long-range rockets. Von Braun helped to lead the Army's efforts. Both camps relied heavily on ideas and systems the Americans had taken from Peenemünde and Mittelwerk.

In addition to employing ex-Nazis to continue former Nazi rocket projects, the Army tried to develop derivatives of the German V-2 *without* directly involving the Peenemünders. The Bumper rocket combined a V-2 as a first stage with a second stage based on a Corporal sounding rocket. The Army launched eight Bumpers starting in 1948 – six of them from White Sands. The seventh and eighth Bumpers were the first rockets to launch from the US military's new 'long-range proving ground' on Cape Canaveral in Florida.

In 1950 von Braun and his team transferred to Redstone Arsenal near Huntsville, Alabama. There, they got to work on behalf of the Army's Ballistic Missile Agency designing a succession of new rockets. Officially, the Redstone and Jupiter rockets, fueled by liquid oxygen and, respectively, ethanol and liquid hydrogen, were weapons. But rockets by their very nature are dual-use. In military hands, a Redstone was a weapon of the apocalypse capable of lobbing a nuclear warhead a distance of hundreds of miles. In the hands of scientists, the very same rocket – minus the warhead – was a vehicle for exploration.

Officially, von Braun's next three rockets after Redstone and Jupiter – Jupiter C, Juno II and Saturn – strictly were space launch vehicles. Ironically, though, it was Redstone itself, the weapon, that ultimately carried the first American into space. Redstone, which like the Army arsenal was named for the red rocks and soil around Huntsville, was little more than a bigger, tougher and more reliable V-2. It's not for no reason that NASA's history division called the 69-foot, 30-ton rocket 'the favorite of the von Braun group'.

With Redstone, the Germans finally realized their original vision for the 46-foot, 14-ton V-2. Redstone, like the V-2, featured a liquid-fuel engine burning alcohol and liquid oxygen. Where the V-2 produced slightly less than 60,000 pounds of thrust, Redstone produced around 75,000 pounds. Redstone could reach a top speed of 3,800 miles per hour before it exhausted its fuel supply. That was 500 miles per hour faster than the V-2's own top speed. Redstone and the V-2 both had all-inertial guidance systems that combined gyroscopes for stabilization with internal computers that calculated the rocket's position and sent steering commands to fins and vanes.

Von Braun delegated much of the work on Redstone to fellow Peenemünder William Mrazek. Chrysler built the prototypes. The first test took place on August 20, 1953 at Cape Canaveral. It, like so many initial tests of new rocket designs, was a total disaster. The rocket crashed to the ground just five miles from the launch site. But Redstone got better. Between August 1953 and November 1958, von Braun's team tested 37 Redstones. Thirteen of them suffered some sort of malfunction. In the midst of this testing in September 1954, von Braun proposed to take the immature Redstone – which the Army still planned to deploy as a delivery vehicle for atomic warheads – and modify it for space travel.

The then 42-year-old von Braun was in a familiar position. Developing a weapon for some army while secretly – and sometimes not so secretly – hoping to hijack that weapon for scientific purposes. Yet he was seemingly unworried that his weapon might, in the meantime, destroy all of civilization. With Redstone, von Braun hoped to continue what he had started at White Sands in October 1946, when he launched a rebuilt V-2 carrying a Navy spectrograph. The spectograph was America's space instrument but it didn't really qualify as a satellite because it didn't orbit Earth. It went up; it came right back down.

To orbit Earth, a craft needs to achieve just enough velocity at a particular angle in order to move forward while also resisting the steady tug of Earth's gravity. To orbit just above the atmosphere – a 'low Earth orbit' – a craft needs to achieve a speed of 17,800 miles per hour. Getting a 75-pound payload up to that speed requires around 28,000 pounds of thrust. A V-2 produced nearly 60,000 pounds of thrust but wasn't an effective space launch vehicle because it was a heavy rocket. Toward the end of its flight, as its propellant was running out, the mostly-empty rocket was dead weight.

It's not for no reason that all effective space launch vehicles since the start of the Space Age in the 1950s have featured more than one stage. The bottom stage of a rocket is its most powerful but also its heaviest. The first stage gives a rocket a tremendous boost, helping to accelerate it from a standstill to thousands of miles per hour and as soon as it's out of fuel, the stage separates and falls back toward Earth, relieving the remaining stages of the burden of its heavy – and now useless – structure.

The second stage can be lighter and less powerful than the first stage. The rocket is already moving at many times the speed of sound and it's shed a lot of dead weight. Thus each successive stage benefits from the hard work, and ultimate self-sacrifice, of the stage before it. Stacking ever-lighter stages is the only method Earth's rocketeers have ever devised for producing enough thrust *in a sufficiently lightweight vehicle* to accelerate a useful payload to a speed adequate to achieve orbit.

In 1954, von Braun crunched the numbers and concluded that he could orbit a satellite with a four-stage space launch vehicle using only existing technology. A Redstone rocket would function as the first stage. The Army and Navy agreed to collaborate on a Redstone-based space launch vehicle – pending approval from the White House, of course. But the Navy at the same time was floating a space-launch project it would pursue on its own, with no Army involvement. The sea service proposed to enhance its Viking and Aerobee suborbital meteorological rockets until they could haul a small payload into space.

In January 1955 President Dwight Eisenhower picked the Navy-only Project Vanguard to launch America's first true satellite. Reportedly, Eisenhower preferred the Vanguard because it was

Bumper 8, the first rocket to launch from Cape Canaveral, Florida, blasts off on July 24, 1950. Bumper combined a V-2 lower stage with a Corporal upper stage. (US Air Force photo)

a scientific instrument, through and through – it had never been a weapon. Thus it answered Sputnik on Sputnik's terms. The Vanguard satellite was a three-pound metal sphere containing two radio transmitters, one with a battery, the other powered by tiny solar cells. The solar cells at the time were groundbreaking technology.

Von Braun was disappointed but he didn't give up. A little over a year later the Army pitched von Braun's new Jupiter-C rocket – another V-2 derivative – as an alternative to Vanguard. The Pentagon not only rejected the offer, it explicitly forbade the Army to make any plans or develop any technology that might undercut the Navy effort.

Demonstrating a stubbornness bordering on petulance, von Braun and the Army continued offering to launch a satellite on one of the

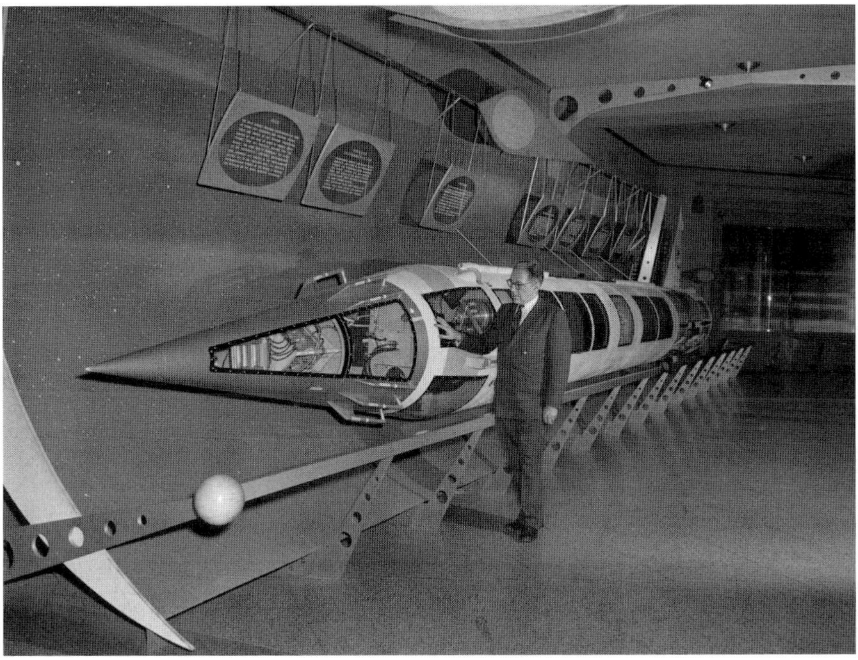

Author and space-advocate Willy Ley poses standing beside a Viking meteorological rocket at the American Museum of Natural History in New York City in 1956. In the post-war period, Ley set out to change Americans' attitudes about space. His 1949 book *The Conquest of Space* signaled a sea change. Ley's book included detailed descriptions of space stations and manned flights to the Moon. (Smithsonian National Air & Space Museum photo)

The US Army's 69-foot, 30-ton Redstone rocket was little more than a bigger, tougher and more reliable V-2. In 1954, Wernher von Braun crunched the numbers and concluded that he could orbit a satellite with a four-stage space launch vehicle using a Redstone as the first stage. (US Army photo)

service's increasingly-sophisticated rockets. In September 1956, von Braun tested a two-stage Redstone that, in the scientist's own estimation, was ideal for a space-launch mission and ready to go *right now*. Again, the Pentagon said no. For the moment in US policy, space belonged to the Navy. But the Navy was slow.

The Vanguard satellite was a three-pound metal sphere containing two radio transmitters, one battery-powered, the other powered by tiny solar cells. The solar cells at the time were groundbreaking technology. Vanguard was so small and light that, when it finally launched in March 1958, it achieved such a high orbit that it could take centuries for Earth's gravity to pull it back down. (Erik (HASH) Hersman photo)

Chapter 14

Sputnik 1

On October 4, 1957, the Soviet Union launched Sputnik 1, Earth's first artificial satellite. It's fair to say the 184-pound, beach-ball-size Sputnik – little more than a metal sphere containing a simple radio transmitter – changed the world.

In response, the (reluctant, Republican) Eisenhower administration and (enthusiastic, Democratic) Congress established the National Aeronautics and Space Administration and began, despite Eisenhower's occasional grousing, budgeting billions and billions of dollars – 4 per cent of federal spending by the mid-1960s – for space science, technology and exploration. All that work culminated in the Apollo 11 mission that, on July 20, 1969, landed astronauts Neil Armstrong and Buzz Aldrin on the Moon while Michael Collins orbited overhead.

Space science had always dovetailed, and forever would dovetail, with *military* uses of space. That was especially true before NASA's establishment. The more money, effort and urgency the government pumped into the nascent US space-exploration effort – then spread out across the US armed services – the more money, effort and urgency also flowed into military space projects.

In that sense, Sputnik 1's 1,440-lap journey around Earth in 1957 and early 1958 both advanced humanity's understanding of its home planet and the cosmos and *also* helped to push humanity to the brink of annihilating itself.

But NASA's founding in response to Sputnik also helped to contain the same demons that rode the Soviet probe into orbit for NASA was a science agency; civilians ran it for civil purposes. As NASA grew and, through bitter bureaucratic battles, began to assume more and more responsibility, the US space program increasingly became a civil pursuit. Yes, NASA's work, in as much as it advanced space science, helped to sustain military initiatives that depended on the same basic research, but the agency's ascendancy also ensured that those military initiatives never went too far.

The United States might develop a space station, but it would be a laboratory rather than a fortress. America might invent a spaceplane, but its purpose would be to rendezvous with orbital science instruments for maintenance purposes and shuttle crews to the aforementioned space station rather than, say, dropping bombs on the Soviet Union. The Soviets, in contrast to the Americans, never actually established a strictly civil space agency. Indeed, the Soviet space program wasn't really a discrete program; it was several *overlapping* and *competing* lines of development, each with a different design bureau as a champion.

There wasn't one Soviet Moon program, there were *two*. One aimed to fly a crew around the Moon, the other planned to land cosmonauts on the lunar surface. Neither succeeded, and it's not hard to attribute that failure in part to the wastefulness of the inter-bureau competition. Soviet leaders and the Soviet press for decades maintained an elaborate fiction that the country's National Academy of Sciences was in charge of the space program. In fact, the dominant design bureau at any given moment tended to be synonymous with the Soviet space program. Through the 1950s and part of the 1960s, Sergei Korolev's Experimental Design Bureau One – that's 'OKB-1' in Cyrillic – was usually the most prominent, as Korolev managed to maintain sole control of most of the USSR's manned space-exploration efforts.

To be clear, the bureaus with their rival chiefs worked for, and at the whim of, the Communist Party. That made them fundamentally political enterprises and in so far as the Party's priorities were military, the Soviet space program's priorities were also military. Government ministers routinely curtailed, or outright rejected, projects that Korolev pitched and which they perceived as being strictly scientific in nature. Of course, if a dramatic science project had obvious propaganda value and thus contributed to national prestige – especially at the Americans' expense – then the ministers were more likely to view it through a military lens and give it the support it needed to have any chance at success.

By the same token, if a space mission or some new piece of space hardware didn't promise some short-term benefit – and, more to the point, a short-term *military* benefit – it faced long odds in the opaque, politically-charged 'system' that encompassed Soviet efforts in space.

Organizationally, the Soviet space program was simultaneously top-heavy, bottom-heavy and personality-driven. It answered to squabbling political leadership and the actual implementation of the leadership's priorities fell to squabbling technical organizations with technicians as managers. What was missing was a healthy, independent and apolitical agency in the middle. One that could weather changes in national leadership and manage, and mediate between, the various technical camps. A Soviet NASA, if you will.

That disparity – between NASA's strong and enduring organization and scientific mission, and the Soviet design bureaus' weakness, volatility and largely military missions – arguably became the space race's defining dynamic. It meant that NASA's ascendence, first over the US military's various space agencies and then over the Soviet design bureaus, represented a triumph of science over war.

It very nearly went a different way. If ever there was potential for one of the Soviet design bureaus to consolidate power and perhaps evolve into an independent agency, it was in the late 1950s during OKB-1's – and Korolev's – peak influence. But it wasn't to be. OKB-1 like any design bureau depended on political patronage. Korolev owed his own successes in part to his close relationship with Nikita Khrushchev, who led the USSR from 1953 to 1964.

Wernher von Braun stands in front of the Saturn V rocket that will eventually boost the Apollo 11 crew on their historic journey to the Moon in July 1969. The Saturn V was the final version of a heavy booster that itself represented the ultimate evolution of the space-launch vehicles von Braun had been developing since joining the US Army's ballistic-missile agency in Huntsville, Alabama in 1950. The Army rockets heavily borrowed from the V-2 rocket that von Braun had designed for the Nazi war effort. In that way, the Saturn V and every space mission it enabled owed their success to Nazi Germany and its campaign of conquest and genocide. (NASA photo)

As long as Khruschev's favor endured, it elevated and protected Korolev and benefited the Soviet space program. 'His stature, and his relationship to Khrushchev, ensured that the political infighting over budgets and priorities were kept to a minimum,' historian Launius said of Korolev.

In the late 1950s and early 1960s Korolev managed to usher through development a host of epoch-making space projects: the R-7 booster, the Sputnik satellites, Space dogs, the Luna probes, Yuri Gagarin's Vostok 1, the first manned mission to punch through the Karman Line and exit Earth's atmosphere, Cosmonaut Alexei Leonov's spacewalk – another first for humanity – as part of the Voskhod 2 mission on March 18, 1965.

But Krushchev's good graces were fickle. As global tensions escalated in 1959 and 1960, Khruschev somewhat soured on Korolev. For the sake of the USSR's position in the world – to say nothing of Khruschev's own standing within the Party – the premier wanted as his ward a scientist who would deliver instruments of power. If not actual weapons, then space systems that clearly demonstrated Soviet might.

From late 1960 until his untimely death less than six years later, Korolev found it much harder to get the resources he needed for his projects – to say nothing of winning approval for new projects. Korolev's slow fall from grace, ending with his death, had profound worldwide implications. 'The lack of focus in the Soviet space program did present some issues, but they did very well until the death of Sergei Korolev,' Launius said.

Chapter 15

Laika the space dog

The Soviets didn't try to hide Sputnik 1 once the satellite was complete and ready to launch. Moscow wanted the world to know it had won the first heat of the space race. On the evening of Sputnik 1's launch, Soviet officials invited reporters into the Soviet embassy in Washington, DC.

The Soviet government, weeks in advance via the country's newspapers, had requested foreigner observers to tune their radio receivers to Sputnik's frequencies between 20 and 40 megahertz. Countless Americans and other foreigners, to say nothing of US and allied intelligence and military agencies, obliged the Soviet request. When Sputnik 1 passed overhead starting on October 4, a lot of people were listening via radio and watching via telescope.

At first, the American public mostly shrugged off Sputnik 1 and the Soviets' follow-on satellite, the 1,100-pound Sputnik 2, which launched on November 3, 1957. A Gallup poll found that only 50 per cent of a sample taken in Washington and Chicago regarded Sputnik 1 as a blow to American prestige. Sixty per cent said that the United States, not the Soviet Union, would make the next great scientific advance.

'Anecdotal material tends to support these figures', a 1958 US government report stated. 'An AP reporter in Sheboygan found that the typical response was a grin and a joke, meaning a refusal to admit that we were falling behind in any way.'

In some government and media circles, by contrast, the reaction to Sputnik 1's orbital *beep beep* quickly escalated to something verging on hysteria. 'US Now Faces Its Gravest Threat,' *The Atlanta Constitution* bellowed. With a Soviet satellite in orbit, the Georgia reporter warned, Americans 'must assume that [the Soviets] have the capability of building intercontinental ballistic missiles and that these can be built in sufficient quantities to launch *en masse* within a few years against targets all over the world.'

Some US lawmakers took a cue from the papers and repeated the dire sentiment right back to them. 'Russia's launching of an Earth satellite, confronts America with a new and terrifying military danger and disastrous blow to prestige abroad,' Richard Russell, a Democratic US senator from Georgia and chairman of the Senate Armed Services Committee, told a *Chicago Daily Tribune* reporter.

There was a clear partisan motive behind Russell's words, of course. Eisenhower was a Republican – and a fairly popular one, at that. Democrats were desperate to exploit any possible weakness in Eisenhower's administration. It's no coincidence that Lyndon Johnson, the Democratic US senator from Texas, the senate's majority leader and a future vice president and president was among the most vociferous critics of Eisenhower's handling of the Sputnik crisis. Johnson launched investigative hearings, during which he repeatedly compared Sputnik 1's launch to the Japanese attack on Pearl Harbor. The main subject of Johnson's hearings was a perceived 'missile gap' between the United States and the Soviet Union. If the Soviets were ahead in space, they were ahead in nukes, too – or so the reasoning went.

Public opinion in the United States, initially sanguine, began to turn with Sputnik 2's November 3, 1957 launch. Sputnik 2 included a pressurized chamber for a passenger, a female dog – a former stray – named Laika. Soviet engineers had designed a harness, food-and-water dispenser and special diaper for Laika in the hope of keeping her alive for a short while. But Sputnik 2 ended up orbiting for five months and, in any event, the craft wasn't designed for safe re-entry.

Laika died after a few hours, possibly from overheating when Sputnik 2's thermal shield failed.

Poor Laika was the subject of much breathless reporting in the United States. *The New York Times* on November 5 devoted several column-inches to the possibility of rescuing Laika in orbit. To the Soviets, Laika was a symbol of sacrifice for the greater good. To many Americans, however, Laika was a symbol of ruthless cruelty on the part of a dangerous enemy. The dog's suffering contributed to the souring of America's national mood.

The Eisenhower administration was desperate to catch up to, and surpass, the cruel Soviets in space but in 1957 there was no NASA and therefore no single American space program. There were *three* space programs, one each belonging to the US Army, US Navy and the US Air Force and in 1957, both the Army and Navy in theory possessed a space-launch system – Explorer and Vanguard, respectively – that was capable of matching the USSR's achievement with Sputnik 1.

As a small-government Republican, Eisenhower opposed creating a new government agency to pick winners and losers between competing space tech and missions. But the logic of a single space agency was obvious as the White House found itself choosing between rival military space projects, as though rockets were political.

'Where should the president place his bets, on Vanguard or Explorer?' recalled Herbert York, who in 1958 headed the new Advanced Research Projects Agency, a small office the Pentagon had established in an effort to mediate interservice rivalries on high-tech programs. York and his staff calculated that both Vanguard and Explorer had a 50-50 chance of failing. Based on that prediction, the White House ordered the Navy to finish and launch Vanguard as soon as possible. The White House *also* instructed von Braun to prepare, as a back-up, his own satellite ... and a Jupiter rocket to orbit it.

In contrast to York's own ambivalence, von Braun was confident the Navy would fail. 'Vanguard will never make it,' he said.

On December 6, 1957, the Navy's 72-foot, three-stage rocket lifted off a launch pad at Cape Canaveral. Vanguard's first two stages burned liquid propellant. The third stage packed solid fuel. It didn't get far. The first stage failed to produce enough thrust. The rocket briefly hovered a couple feet off the pad then settled back to the ground – and exploded. The blast tossed the satellite clear of the flaming wreckage. Its radios were still beeping when technicians pried it open. The press dubbed the abortive launch 'Flopnik' and 'Failnik'.

Four months later, the Vanguard team succeeded in orbiting one of its tiny satellites. But by then, Vanguard was an historical afterthought. The Vanguard 1 spacecraft was so small, and achieved such a high orbit, that it could take centuries for Earth's gravity to pull it back down. In 2021 its radios are dead, but it's still up there, six decades after blasting off – Earth's oldest satellite and, as the heir to disgrace, one of its most disappointing.

The Navy's failure in December was, for von Braun and the Army, the opportunity of a lifetime. Two months later on January 31, 1958, von Braun's Jupiter rocket sat on a pad at Cape Canaveral as Army controllers counted down to launch. Von Braun wasn't in Florida on this historic day. Travel delays had stranded him in Washington, DC so he gathered with Jet Propulsion Laboratory director William Pickering, satellite-designer James Van Allen and a coterie of Army officers and officials in a communications room inside the Pentagon. 'Somewhat studiously calm, perhaps, but confident,' is how one attendee described von Braun.

They listened as a radio news program relayed updates from a reporter standing miles away from the launch pad, watching the rocket through binoculars. At the same time, the controllers at Cape Canaveral sent out official updates via teletype.

At 10:48 pm the rocket's main stage ignited. 'Slow rise, faster, faster!' the radio reporter crowed. 'Straight up, found a hole in the clouds! Is beginning to tilt on course. Is apparently still on course, rising and going away. It is out of sight, but it must be successful!'

The teletype was more sanguine, simply repeating 'still going' every 10 or 20 seconds, finally noting the rocket passing through the jet stream. Several tense minutes passed as a computer on the ground calculated the precise seconds to send radio signals instructing the rocket to separate first one stage, then another and another.

In those early days of space exploration, there weren't a lot of ways to track a satellite. The Army was counting on crews at listening stations on the ground to pick up radio signals from Explorer 1 in order to confirm that the satellite was in the planned orbit and

The R-7 was the Soviet Union's first intercontinental ballistic missile and space-launch vehicle. Owing to the substantial bulk of the USSR's atomic warheads compared to US warheads, lead designer Sergei Korolev had no choice but to scale up the R-7 to a huge size relative to American ICBMs. The R-7 quickly became obsolete as an ICBM, but its excess of power – and thus payload – made it highly successful as a launch vehicle. An R-7 boosted the Sputnik satellites into orbit in October 1957. Versions of the R-7 were still in use in 2021. (RKK Energia photo)

functioning as designed. Confirmation took more than an hour. When three West Coast stations reported signals, von Braun knew he and his team had succeeded. After a brief celebration in the Pentagon, von Braun, Pickering and Van Allen hurried into the district for a 1:00 am press conference.

Someone wondered if any press would show. It was late, after all, it wasn't clear just how much the media – to say nothing of everyday Americans – cared after the farce of Vanguard. But the room was packed. Von Braun, Pickering and Van Allen held aloft a scale model of their winning rocket and satellite. Granted, it was a third-place winner after Sputnik 1 and Sputnik 2. Still, it was a triumph just to be in the race.

Analysis later confirmed that Jupiter performed flawlessly. The 30-pound Explorer 1 satellite separated at an altitude of 900 miles. It would orbit Earth 58,000 times before re-entering the atmosphere and burning up in 1970.

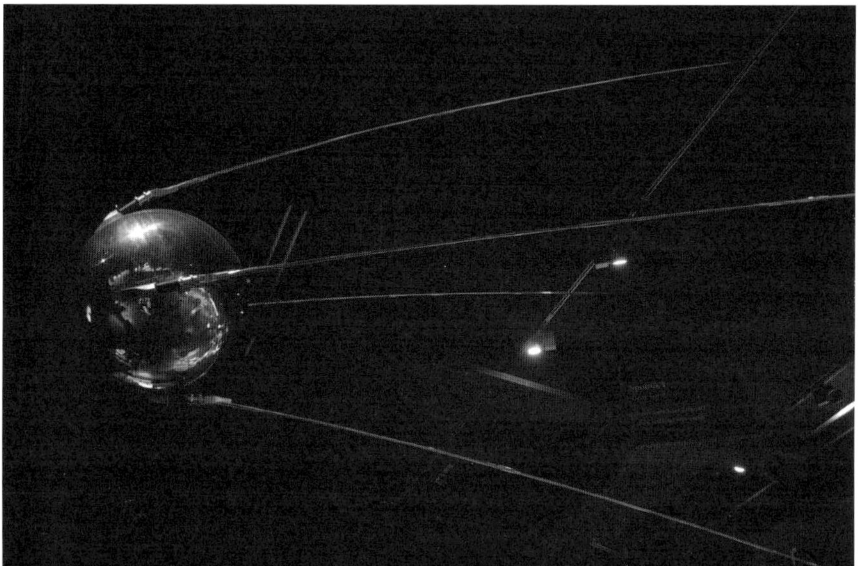

The 184-pound, beach-ball-sized Sputnik 1 was little more than a metal sphere containing a simple radio transmitter but it changed the world. This reproduction of Sputnik 1 is on display at the Deutsches Museum in Munich. (Photo by Flickr-user Killfile)

On December 6, 1957, the Navy's 72-foot, three-stage Vanguard rocket – America's first attempt at a space launch vehicle – lifted off at Cape Canaveral in Florida. The rocket briefly hovered a couple feet off the launch pad then settled back to the ground and exploded. The blast tossed the grapefruit-size satellite clear of the flaming wreckage. Its radios were still beeping when technicians pried it open. (NASA photo)

Jet Propulsion Laboratory director William Pickering, scientist and satellite-designer James Van Allen and Wernher von Braun hold a model of von Braun's Explorer 1 rocket on January 31, 1958 – two hours after the rocket successfully orbited America's first satellite. (NASA photo)

Chapter 16

The Vanguard disaster

In the aftermath of the Vanguard disaster, Eisenhower's approval rating dropped 22 points. He belatedly addressed the country in August 1958. With only also-ran satellites of his own to celebrate, he tried to downplay the significance of the Sputniks to the military balance of power. A launch vehicle didn't necessarily have strategic military applications, Eisenhower said. Just because the USSR could shove a 184-pound metal ball into orbit didn't mean it could lob an atomic warhead across the Atlantic Ocean to flatten an American city.

Nor did a 184-pound metal ball, or a dog in an orbiting kennel, portend some all-seeing Soviet eye in outer space. 'I think that given time, satellites will be able to transmit to the Earth some kind of information with respect to what they see on the Earth,' Eisenhower said. 'But I think that period is a long way off.' Eisenhower was wrong about the likely capabilities of ever-more-sophisticated satellites, but he was correct to downplay the Soviets' achievement in tossing Sputnik 1 and Sputnik 2 into orbit.

Yes, the Soviet Union had beaten the United States into space, but that first-place finish belied the USSR's struggle in the middle stretch. To outside observers, it appeared that the Soviets dramatically succeeded in their very first attempt to launch a satellite. In fact, three Soviet launches failed in the summer and fall of 1957 before the Sputnik 1 mission finally got lucky. Moscow managed, by way of a concerted campaign of censorship targeting newspapers and scientific journals, to cover up those failures.

In the late 1950s, the Soviet and American space programs were equally diffuse, inexperienced and failure-prone. They were both developing roughly similar systems at roughly the same pace – and for good reason, both depended on Nazi expertise. But where von Braun on the American side was a public figure, the Germans who labored on behalf of the Soviet space program did so in secrecy. The Germans working for the Soviet Union not only weren't famous, their names have largely disappeared from history.

At the same time the Americans were resettling von Braun and his team in the United States under Operation Paperclip, Soviet officials under wartime armaments commissar Dmitry Ustinov were rounding up their own ex-Nazi scientists and engineers. The effort began with a secret decree from the Soviet Council of Ministers, which competed with the Central Committee of the Communist Party for overall leadership of the USSR. The May 13, 1946 decree was signed by Ustinov and Georgy Malenkov, a future premier.

'The following work on reactive technology in Germany shall be considered top-priority tasks,' the decree stated, using a Soviet euphemism for special weaponry. The tasks included 'the complete restoration of the technical documentation and models of the V-2 long-range guided missile.'

A 'Reactive Technology Committee' would form, with the mission of 'selecting the necessary number of specialists with various backgrounds from the corresponding ministries and sending them to Germany to study and work on reactive armaments, keeping in mind that each German specialist shall be assigned a group of Soviet specialists so that the latter may gain experience. The transfer of the design bureaus and German specialists from Germany to the USSR by the end of 1946 shall be predetermined', the decree added. It also approved high salaries for any German rocket experts the Soviet committee recruited.

The Soviets' pet Germans actually stayed in Germany for a short while. The Soviet Union actually had a fairly successful rocket program as early as 1936, but it all but collapsed during Soviet premier Joseph Stalin's paranoid purge of officials he suspected of being disloyal. Korolev and Valentin Glushko, the top Soviet rocket engineers, were arrested and sent to work camps. But the V-2 startled Stalin. In 1944, the dictator released Korolev and Glushko. Korolev in particular was in bad shape. After months laboring in the Kolyma gold mines, the scientist was missing most of his teeth and suffering serious kidney and heart problems. He would die in 1966 at the young age of 59.

Korolev got to work piecing together the fragments of the Soviet rocket program. There wasn't much left, and in the fall of 1945 the government made the surprising decision to relocate Korolev and his fellow scientists and engineers to war-ravaged Germany. They settled into two former V-2 sites and, together with thousands of von Braun's former workers, began the hard work of recovering documentation, tooling and components for the production of V-2s.

The Soviets didn't stay long. In the spring of 1946, Moscow recalled the Soviet rocket-developers. That fall, without warning, Moscow's agents began rounding up more than 2,000 German V-2 workers and their families and, under Operation Osoaviakhim – the Soviet answer to Paperclip – herded them into rail cars.

The Germans eventually arrived at a secret research facility – a former anthrax research site – at Gorodomlya Island in central Russia. The whole rocket effort fell under a new government body, Scientific-Research Institute No. 88.

In the meantime, work resumed on the Soviet V-2 at a former air force airstrip near the Podlipki train station in Kaliningrad, a suburb of Moscow. Over time, Podlipki would become the home of the Soviet, then Russian, space program. In 1996 Russian authorities renamed it Korolev City. It took more than a year for the Soviet-led, German-staffed team in Podlipki, mostly working out of specially-equipped railroad cars, to assemble a V-2 using leftover German components. On October 18, 1947, the V-2 blasted off from Kapustin

Yar, a military test site near Moscow. The rocket flew 128 miles but strayed nearly 19 miles to the left and disintegrated before impacting the test range.

The next launch, two days later, was equally disastrous. Once again, the rocket deviated to the left of its plotted course. 'It went toward Saratov,' a range official wryly commented. Saratov is a town more than 200 miles north of Kapustin Yar. Officials overseeing the rocket program called an emergency meeting. 'Imagine what would have happened if the rocket had reached Saratov,' state security general Ivan Serov, future head of the KGB, told rocket-engineer Boris Chertok. 'I won't even begin to tell you; you can guess what would have happened with all of you.'

Chertok and his fellow rocketeers weren't overly worried. The V-2 couldn't reach Saratov from Kapustin Yar. They crunched the numbers and determined that rocket number-two flew 144 miles but strayed a staggering 112 miles to the left. Something was wrong. Ustinov asked the Germans to figure it out – a move Chertok later described as 'annoying'. Two German scientists, Kurt Magnus and Hans Höch, had been sitting around the test range 'without anything in particular to do,' according to Chertok. 'This is your rocket and your instruments; go figure it out,' Ustinov told the Germans. 'Our specialists don't understand why it went so far off course.'

Magnus and Höch sat down in a laboratory attached to a train supporting the test program. They discovered that vibrations could interfere with the connection between the rocket's internal gyroscope and its control surfaces. The solution was simple. Höch grabbed some spare parts from the railroad car and cobbled together a filter that blocked the disruptive signal resulting from the interference. The next V-2 the Soviet-German team launched stayed on course.

The Nazis actually had the same problem during the war – and it had caused countless V-2s to wander. 'If Dr. Magnus and Dr. Höch had worked with us in Peenemünde during the war, British losses during our bombardment of London would have been considerably greater,' joked Helmut Gröttrup, a former launch officer in the Nazi rocket force.

In all, Korolev's team launched 11 V-2s they assembled from German parts. Five impacted their target areas, making them roughly as reliable as wartime V-2s, according to Chertok. 'The flight tests in 1947 showed that Soviet specialists, both military and civilian, had mastered the fundamentals of practical rocket technology and had gained the experience needed to make an accelerated transition to a now independent development of this new, promising field of human endeavor,' Chertok noted. In other words, the Soviets had gotten what they needed from the ex-Nazis. Moscow sent its Peenemünders home. America, however, held on to *its* Peenemünders.

The advantage of the US policy soon was clear. After those 11 V-2 launches, the Soviet rocket program moved on to its next goal. 'We immediately switched over to the task of developing domestic rockets,' Chertok wrote. The Soviets called their V-2 clone the 'R-1'. Korolev's team had to recreate, or replace, all the alloys and other materials the Germans used to make the V-2. It was a big undertaking. 'It was necessary to select 86 grades of steel, 56 grades of nonferrous metals, 159 nonmetallic materials,' explained RKK Energia, a Russian rocket company that grew out of Korolev's work in the 1940s.

Copying the internal systems proved equally tricky. Of the first nine R-1s the Soviets launched in tests, just one – in October 1948 – struck its target zone. 'Causes of the accidents were mainly assigned to a low manufacturing quality of the rocket assemblies and systems, an insufficient amount of checkout testing of units and instruments and immaturity of certain systems,' RKK Energia stated.

Progress was painful but steady. The next batch of 20 rockets was more reliable. Seventeen struck their target zones. Moscow was finally happy and the R-1 entered front-line military service in November 1950.

In the same way that von Braun had donated V-2s to his physicist colleague Regener for meteorological research during the war, Soviet scientists piggybacked on R-1 testing to get weather sensors into the upper atmosphere. After Korolev's team had finished the first military R-1 model, it began work on strictly scientific V-2 clones. The main difference, aside from the removal of the warhead, was that the civilian V-2s flew a tight, vertical trajectory that maximized their altitude performance. By contrast, the military R-1 – like the V-2 – flew a flat trajectory that traded height for distance.

On August 15, 1951, an R-1 belonging to the Soviet Academy of Sciences, modified with a special pressurized compartment, carried two female dogs – Dezik and Tsygan – to an altitude of 68 miles. The compartment separated from the booster, deployed a parachute and safely floated back to Earth. Dezik and Tsygan were unharmed. A few

The Soviets launched their first captured V-2 from Kapustin Yar on October 18, 1947. The rocket flew 128 miles but strayed nearly 19 miles to the left and disintegrated before impacting the test range. The next launch, two days later, was equally disastrous. 'It went toward Saratov,' 200 miles to the north, a range official wryly commented. (Russian Ministry of Defense photo)

weeks later, Dezik blasted off on her second mission along with a new partner, Lisa. Tragically, the parachute failed, sending the pups plummeting to their deaths. Korolev and his team were reportedly devastated. After Dezik died, physicist Anatoli Blagonravov adopted Tsygan as a pet. But the animals' sacrifice was not in vain. The Soviet Union had proved it could boost a living being past the edge of space and *usually* bring it safely home.

The Soviet space program used dogs and other animals to test out crew capsules and other systems before risking people in them. Belka and Strelka – pictured here with scientist Oleg Gazenko, a leader in the space-dog initiative – were the first dogs to reach space and return safely, in August 1960. (Soviet Institute of Biomedical Problems photo)

Arrested for disloyalty in 1938, Sergei Korolev – pictured here with cosmonaut Yuri Gagarin – spent months laboring in the Kolyma gold mines. He was missing most of his teeth and suffering serious kidney and heart problems by the time Soviet premier Joseph Stalin released him in 1944. Korolev rose to head Experimental Design Bureau One, the core of the Soviet space program. He would die in 1966 at the age of 59, leaving unfinished his most important project – the N-1 Moon rocket. (RKK Energia photo)

Chapter 17

Nuke the Moon

The Soviets followed up the Sputniks with a series of Moon probes called 'Luna'. Actually, the term 'probe' might be too generous. To twenty-first century readers, the word implies a sophisticated scientific instrument requiring precise control and delicate handling. Picture a wheeled rover descending toward the surface of Mars underneath a hovering crane vehicle. To Soviet planners in 1958 and '59, probing the Moon meant packing 800 pounds of gas-detectors, cosmic ray readers and radio telemetry gear inside a 2,400-pound metal sphere with several antennae protruding from it, piling it on top of a three-stage version of the R-7 rocket and firing directly at the Moon.

A Luna probe wasn't supposed to land on the Moon. It would fly straight into the giant rock in a catastrophic crash of dust and metal. The only data the probe would return to Earth would be the data it gathered and transmitted before its fatal collision. For all its brute simplicity, Luna nevertheless would represent a significant advancement. In the first lap of the space race, the Soviets and Americans competed to be the first country to launch a satellite into Earth orbit. In the second lap, the goal was to boost a manmade object free of Earth's gravity.

It wasn't terribly important to either country what the probe did once it escaped Earth's embrace. Neither was either country concerned with the probe's long-term survival. *The escape was the point.*

As with Sputnik, the Soviets got there first. After three failed launches, on January 2, 1959 Luna 1 successfully lifted off from Baikonur and, 12 minutes later, achieved the seven-miles-per-second velocity it needed to shrug off Earth's gravity. But then the probe's telemetry failed and instead of hitting the Moon, it flew wide and settled into a 90-by-122-million-mile orbit around the Sun, effectively making it a tiny, man-made planet between Earth and Mars.

The Americans chased after Luna 1 with their own Moon probe. The US Army's Pioneer 4 launched from Cape Canaveral atop a Juno rocket on March 3, 1959. Von Braun and his colleagues at the Jet Propulsion Laboratory meant for Pioneer 4 to pass close to the Moon and beam back photos but a timing error aboard the rocket flung Pioneer 4 past the Moon at a distance of 37,000 miles instead of the planned 20,000. That was too far away for the probe's sensor to work.

The Soviets once again had beaten the Americans in space. In their humiliation, the Americans lashed out, threatening to nuke the Moon just to prove to everyone else that they could.

Leonard Reiffel was a physicist at the Illinois Institute of Technology when, around May 1958, the New Mexico-based Air Force Special Weapons Center came to him with a request. 'We were asked by the US Air Force to add a small, fast-track project to investigate the visibility and effects of a hypothetical nuclear explosion on the Moon,' Reiffel recalled years later. 'I was told the Air Force was very interested in the possibility of a surprise demonstration explosion, with all its obvious implications for public relations and the Cold War.'

Reiffel assembled a team that included a young researcher named Carl Sagan, who in coming years would become America's – perhaps the world's – most famous astronomer. The following June, they submitted their report.

'The motivation for such a detonation is clearly threefold: scientific, military and political,' Reiffel and his team wrote. 'The scientific information which might be obtained from such detonations is one of the major subjects of inquiry of the present work. On the other hand,' they added, 'it is quite clear that certain military objectives would be served since information would be supplied concerning the environment of space, concerning detection of nuclear device testing in space and concerning the capability of nuclear weapons for space warfare. The political motivations for and against the detonation of a nuclear weapon are equally clear and are, in reality, outside the scope of the present work,' Reiffel and his team continued.

'Obviously, however, specific positive effects would accrue to the nation first performing such a feat as a demonstration of advanced technological capability. It is also certain that, unless the climate of world opinion were well-prepared in advance, a considerable negative reaction could be stimulated. Obstacles to detonation of a nuclear weapon on the Moon, from a scientific viewpoint, center around environmental disturbance, biological contamination and radiological contamination, only the last of which is unique to the nuclear weapon.'

The report's dry language belied the scientists' actual horror at even being asked to consider the effects of nuking the Moon for what were, they suspected, purely propagandistic purposes.

'Whether the project was motivated by a desire for the United States to impress the world (and the Soviet Union in particular), or by fear that the Soviet Union itself might try the stunt, I cannot say,' Reiffel wrote in 2000. 'It was emphasized, however, that the most sensitive aspect of the project was, as with many other Department of Defense projects, its very existence. Our task was to assess what gross visible and other phenomena might be generated by the explosive release on the lunar surface of a given number of kilotons of energy. We were also asked to try to determine what legitimate scientific data might be gathered from such an event: for example, about lunar chemistry. The cost to science of destroying the pristine lunar environment did not seem of concern to our sponsors – but it certainly was to us, as I made clear at the time.'

Reiffel tasked the then 25-year-old Sagan to model the expansion of the cloud of gas and dust that might result from an atomic explosion on the Moon. The modeling work was central to the Air Force project. After all, the military wanted the world, and the Soviets in particular, to *see* the lunar mushroom cloud. To appreciate that, yes, the USSR might have beaten the USA into orbit and further into the solar system, but the United States still possessed scientific and military power to spare. America might be second to probe the Moon, but it could be first to blow it up.

Sagan struggled to determine, with any certainty, whether people on Earth would be able to see a nuclear blast on the Moon without powerful telescopes, plenty of advance notice and highly favorable conditions. Sagan's nuke-the-Moon research very nearly got him in a lot of trouble. Despite signing a non-disclosure agreement and being subject to government secrecy rules, the young astronomer reportedly circulated a manuscript describing his work on the Air Force Moon demonstration. He also mentioned the nuke plan in an application he submitted for a fellowship in 1959.

Reiffel didn't become aware of Sagan's disclosures until decades later. 'In my opinion Sagan breached security,' he wrote.

Back in 1958, Reiffel brought in additional researchers to help Sagan. Even working together, they couldn't say for sure whether nuking the Moon would work even within its wrongheaded, military-dictated parameters.

'Since the Moon is not of uniform luminance, the results are not rigorous,' Reiffel and his team wrote as they presented their data. 'Ranging from unresolvable points to large irregular areas and from bright spots in deep shadows, arranged in most irregular fashion, the Moon represents a difficult background in which to search for a small area.'

'Fortunately for the future of lunar science, a one- or two-horse race to detonate a nuclear explosion never occurred,' Reiffel later wrote.

It's not clear why the Air Force demurred, but Sagan and the others' inability to say for sure whether the blast even would be clearly visible on Earth surely was a factor. What's the point in blowing up the Moon to scare the Soviets if the Soviets might not see the blast? Even if it wasn't an evil idea, it would have been a bad one. In retrospect, the abortive military effort to nuke the Moon was a compelling argument in favor of civilian control of a science-focused US space program.

Wernher von Braun, Jet Propulsion Laboratory's John Casani and James Van Allen, a University of Iowa physicist, inspect components of the Pioneer 4 spacecraft on March 1, 1959. A timing error aboard the rocket flung Pioneer 4 past the Moon at a distance of 37,000 miles instead of the planned 20,000. That was too far away for the probe's sensor to work. (NASA photo)

PART THREE

VERY HIGH

Space-planes, space stations,
one very bad movie
… and the Moon

Chapter 18

Man Very High

Having placed America's first satellite in orbit atop one of his evolved V-2s, von Braun's next step was obvious. While the Air Force squandered (some of) its time and money on an outrageous scheme to nuke the Moon, the Army would take one of its new launch vehicles, stick a man on top, blast him nearly into orbit and, hopefully, blow right past the Soviets.

In February 1958, von Braun floated an idea he called Man Very High. For $12 million, he told his bosses, he would boost an astronaut right up to the Karman Line – 'space'– if you're being generous. The whole project would take less than two years, von Braun claimed. The idea was a kluge of existing initiatives. Von Braun planned to take his reliable Redstone rocket and pack into its nose a 1,700-pound enclosed metal gondola he would borrow from the Air Force's Man High balloon project, which ultimately carried men to an altitude of 100,000 feet three times in 1957 and 1958.

The Redstone would lob the gondola to a peak altitude just below the Karman Line. Arcing high over Earth, the man in his metal closet would experience six minutes of weightlessness before splashing down into the ocean a couple hundred miles from the launch pad. The Navy would position a ship to retrieve the astronaut.

Man Very High was a desperate, Frankenstein's monster of a project but there was a reason for von Braun's urgency. In addition to plotting an absurd atomic attack on the Moon, the Air Force also was organizing a much more viable – which is not to say *actually viable* – project.

Man In Space Soonest aimed to boost a human being through the Karman Line by late 1960.

Von Braun's proposal didn't get very far at first. The Air Force for obvious reasons declined to participate, not wanting to undermine its own, desperate orbital push. The Army repackaged the effort, renamed it Project Adam, and pitched it to the Pentagon's Advanced Research Projects Agency.

Perhaps recalling the period of his youth when he'd hitched his scientific fortunes to the Nazi war machine, von Braun tried to sweeten the Project Adam proposal. The same rocket-and-capsule combination that could send an astronaut along the Karman Line also could transport a small number of soldiers very quickly across a great distance, albeit at eye-watering expense.

The ARPA managers scoffed. So did Hugh Dryden, director of the National Advisory Committee for Aeronautics, which in just a few months would become the National Aeronautics and Space Administration. Dryden told a House of Representatives committee that 'tossing a man up in the air and letting him come back ... is about the same technical value as the circus stunt of shooting a young lady from a cannon'.

The secretary of the Army went to bat for von Braun's scheme. The technical and military merits of Project Adam shouldn't be the only metrics by which authorities judged the proposal, Wilbur Brucker argued. He described Project Adam as a possible 'national political psychological demonstration'. In other words, propaganda in the same vein as the nuke-the-Moon initiative. The Soviets had beaten the Americans into orbit with unmanned satellites. Brucker argued it would benefit the United States to beat the Soviet Union into space with a manned spacecraft – even if that spacecraft was crude. Sometimes being first matters.

Owing to relentless interservice rivalry and the duplication of space efforts across military and civil agencies, the US government had options, however. There was the Air Force man-in-space program, of course and the Navy had just proposed its own manned spacecraft, as well – the Manned Earth Reconnaissance system. MER was weird. It relied on a whole lot of new and unproven technology. The crew capsule launched atop a two-stage booster, after separating, the cylindrical capsule elongated by way of telescoping beams and sprouted wings, transforming into a delta-wing glider for its descent back to Earth.

While the Army, Air Force and Navy lobbied for their respective – and wildly different – manned space projects, the embattled president reluctantly backed the legislative effort to expand NACA and give it authority over space exploration, including manned spaceflight.

America's space program was in flux. Until it settled, Defense Department brass were reluctant to back one manned-spaceflight effort at the expense of any other. The Pentagon kicked von Braun's Project Adam back down to the Army for further study. Thus began a period of delay, confusion and conflict that, once again, gave the Soviets under Korolev a chance to pull ahead in space.

On July 29, 1958, on the advice of Vice President Richard Nixon and science advisor James Killian and with the Democratic Congress overwhelmingly in favor, Eisenhower signed the National Aeronautics and Space Act of 1958, creating NASA from the raw materials of the existing NACA. Dryden stayed on as deputy under the new space agency's first administrator, T. Keith Glennan.

The Space Act made NASA – not the Army, Navy or Air Force – the lead agency for America's exploration of space. Among the new agency's missions, according to the law, were 'the development and operation of vehicles capable of carrying instruments, equipment, supplies and living organisms through space'. 'Living organisms' included human beings. Officially though not yet exclusively, NASA was in charge of human spaceflight.

The division of labor at first was unclear. It took a whole year of negotiation for the Pentagon and NASA to agree to boundaries. In the meantime, work got underway on a rocket that, more than any other single piece of hardware, would become von Braun's legacy in the United States. In September 1958, the Army and the Advanced Research Projects Agency tasked von Braun with developing a superheavy booster that, the agencies hoped, would solve America's biggest space-lift problems. It was one thing to punch a small satellite or a one-man capsule through the Karman Line into orbit around Earth. It was quite another to shoot a probe or capsule with several crew free of the planet's gravity and send them hurtling toward the Moon or beyond.

Eying heavier payloads and further-traveling space missions, von Braun had been tinkering with a booster concept that involved strapping eight Redstones to a Jupiter. He called his new liquid-fuel rocket 'Saturn' – as in the planet *beyond* Jupiter. It was an ambitious design. In its ultimate form, the Saturn V, it would generate a thundering 7.6 million pounds of thrust, making it humanity's most powerful rocket by far – a record it still holds in mid-2021. And it was reliable. NASA launched 32 Saturn rockets between 1961 and 1975. One Saturn V suffered a partial failure of the engines in its second and third stages during a 1968 test flight, but even that mission still reached orbit and none of the Saturns crashed.

The huge rocket in its various models took the Apollo astronauts to the Moon. Saturns also boosted the components of the Skylab space station and ferried crews to the station. In the rocket's final act, it carried an Apollo capsule to an orbital rendezvous with a Soviet Soyuz capsule on July 17, 1975 – the first cooperative space mission between rival superpowers.

Saturn, the rocket that carried Americans to the Moon, almost had an explicitly military mission. Today, however, Saturn's legacy is a peaceful one. Its success helped to obscure the death, destruction, slavery and exploitation that, without Saturn, might be engraved on von Braun's tombstone.

Before it transported a single probe or manned capsule through the Karman Line, Saturn transported von Braun on a career-defining journey between US government agencies. A presidential memorandum dated October 21, 1959 directed NASA to absorb von Braun's work at Redstone as well as the Army's Jet Propulsion Laboratory, a key ally of von Braun's team. Not everyone in the new space agency wanted him. Robert Gilruth, who oversaw an early study group inside NASA and soon rose to head the agency's mission-control center in Houston, was particularly wary of the German. 'Von Braun doesn't care what flag he fights for,' Gilruth commented.

The final shove that ushered von Braun out of the Army and into NASA ironically came from a top military bureaucrat. Now the Pentagon's top research official, former ARPA chief Herbert York was determined to end, once and for all, the seemingly endless squabbling between the armed services – the Army and Air Force, in particular – over roles and responsibilities in space. 'My worst problem was the Army because von Braun was in the Army,' York recalled years later. He sensed that, if he managed to transfer Saturn – and thus von Braun – into NASA, it would gut the Army's space program and reduce by at least one the number of military agencies with a claim on Earth's orbit.

York did not like von Braun, but he believed he understood the German. 'Some people regard von Braun's unwavering dedication to the grand dream of space flight as heroic and farsighted,' York wrote in his 1987 autobiography *Making Weapons, Talking Peace*. 'Others cannot overlook the grotesque means and unprincipled behavior he used to realize his dreams. I am among the latter, but in this instance I was glad to exploit his willingness to go, without argument, wherever the money was.'

York's strategy was a clever one. He formally questioned the rising cost of developing Saturn. This implied softening military support for the heavy booster. Von Braun didn't fail to notice that NASA's own interest in Saturn never flagged. 'All we want is a very rich and benevolent uncle,' he told Glennan, only half-jokingly.

The Army didn't just give up von Braun without a fight, however. In late October 1959, Glennan traveled to Huntsville to arrange the transfer of the Redstone and JPL programs. He ran into rooms full of irate officers. One general flatly refused to cooperate. He accused Glennan of 'breaking up the von Braun team', Glennan wrote in his diary. 'To an extent, this was the proper characterization, in all truth,' Glennan wrote. 'I had not realized at that time how much of a pet of the Army's was von Braun and his operation. He was their one avenue to possible fame in the space business. I left the meeting with my tail between my legs,' Glennan recalled. NASA swiftly subsumed a number of military space projects including the Air Force's Man In Space Soonest. But it would be another year before the Army finally relinquished Redstone and JPL.

NASA named von Braun the director of the Marshall Space Flight Center and all but handed him the keys to the US space program. Marshall was just one of NASA's eventual ten research and test centers, but under von Braun MSFC grew to be far and away the biggest. At its peak in 1966, Marshall employed around 7,700 staff, one of every five NASA workers. Its annual budget swelled to a billion dollars – that's $9 billion in 2021.

NACA – later, NASA – had been working on human spaceflight before von Braun came over from the Army. The agency had announced its first seven astronaut candidates for Project Mercury back in April 1959, but NASA's man-in-space efforts got a huge boost from von Braun and Redstone's transfer into the agency. NASA not only wanted Saturn, it also had plans for von Braun's Project Adam rockets, which were suitable for manned, suborbital tests of the Mercury hardware. If those tests were successful, NASA could slap the same gear on a more powerful Atlas rocket and send a man through the Karman Line – hopefully before the Soviets did. Meanwhile, the agency would wait on Saturn for missions beyond Earth's orbit.

The first Mercury-Redstone test, at Cape Canaveral on December 19, 1960, was messy. High winds delayed the morning launch, then a valve failed, resulting in further delays. The rocket, fueled by liquid-oxygen and ethanol, finally launched, carrying an unmanned capsule 131 miles into the air. The booster and capsule separated and began their individual journeys back down to Earth. The Navy steamed in to retrieve the capsule. So far, so good.

But then von Braun's people analyzed the data and developed the film from the movie camera inside the capsule. The film revealed evidence of sloppy workmanship – washers, nuts and wire clippings floating inside the capsule.

For its part, the data showed that, on re-entry, the capsule achieved an extra G of acceleration above what the team had projected. That extra G was enough to begin knocking out the capsule's occupants. Engineers would need to alter the flight profile for future missions if they wanted the crew to stay conscious.

The 'crew' for the next Mercury-Redstone mission was a chimpanzee named Ham. The four-year-old chimp beat out five other chimps for the dubious honor of taking the Mercury capsule for a suborbital check-ride. NASA trained Ham to pull functionless levers – roughly simulating the operation of a spacecraft – while staying calm in a small space undergoing radical changes in temperature and acceleration. His training involved punishment and reward. Fail to pull a lever, and he got an electrical shock. Pull the lever as required, and he got a banana treat.

Luckily for Ham, the launch on January 31, 1961 was only a minor disaster. The capsule once again racked up more Gs than the planners had projected. The force of acceleration was supposed to peak at 12gs. Poor Ham actually endured nearly 15 times normal gravity while pulling his fake levers to avoid a shock. Gemini's designers hadn't anticipated the effects waves might have on a capsule after a water-landing. It took three hours for Navy crews to reach Ham in his capsule. By then, the impact of the waves had pried open the spacecraft and nearly half a ton of seawater was sloshing around inside. But Ham survived and when they finally pulled him out of the capsule, he happily gobbled an apple and half an orange.

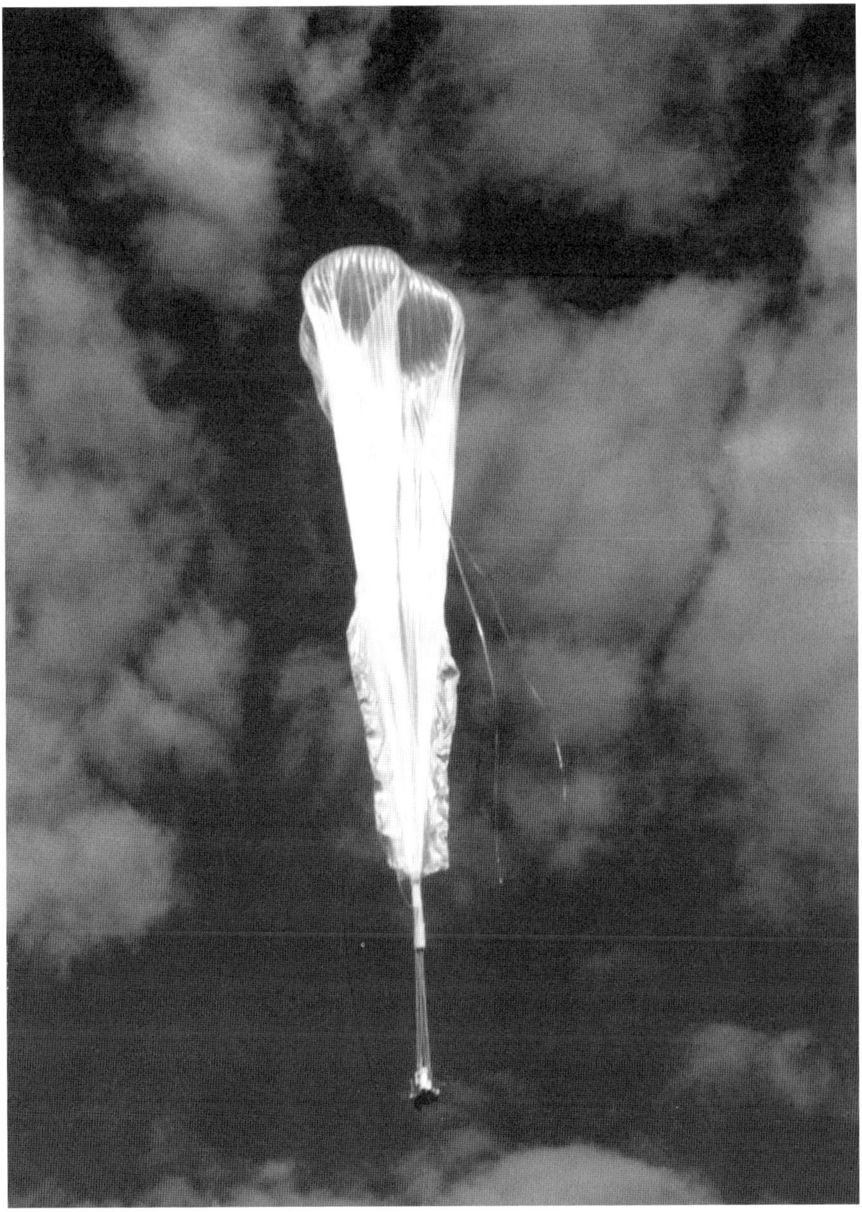

In August 1957, US Air Force Major David Simons ascended to 101,516 feet in one of the Air Force's Man High balloon gondolas. The ascent and descent over Minnesota took 32 hours. Wernher von Braun proposed to borrow the Man High gondola for use as the crew capsule in a manned spacecraft boosted by a Redstone rocket. (US Air Force photo)

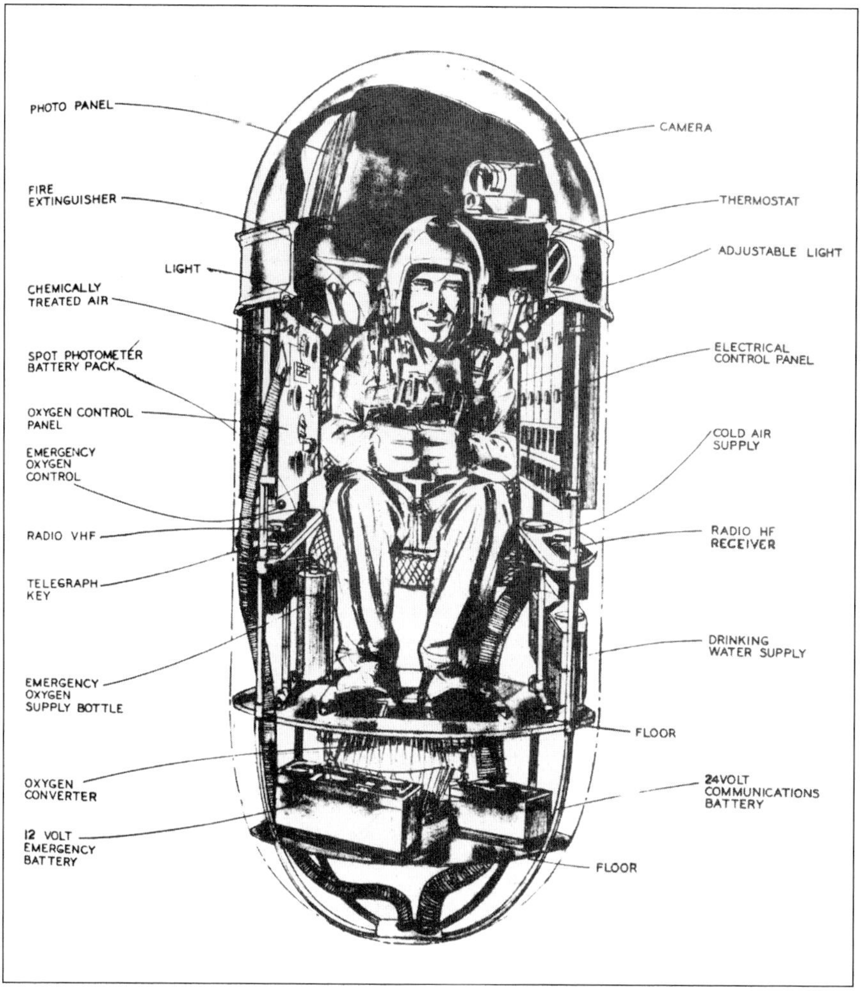

Above: The US Air Force's high-altitude Man High balloon gondola was anything but comfortable, but crew comfort was low on Wernher von Braun's list of priorities when he proposed adapting the gondola for use as a manned spacecraft. Hugh Dryden, director of the National Advisory Committee for Aeronautics, compared von Braun's proposal to 'the circus stunt of shooting a young lady from a cannon'. (US Air Force art)

Opposite above: Sleepless and alternately very cold and very hot as he rode a Man High balloon to 101,000 feet, US Air Force Major David Simons nevertheless made important observations about human physiology in a near-space environment. He snapped this self-portrait as he approached the record altitude. (US Air Force photo)

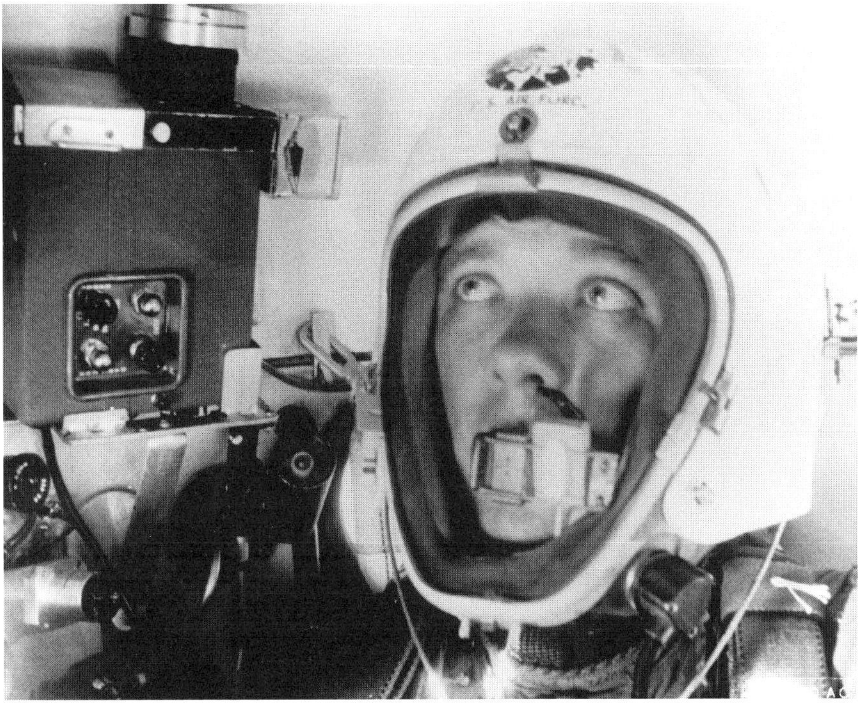

The US Defense Department rejected Wernher von Braun's proposal to modify the US Air Force's Man High balloon gondola for use as a manned spacecraft but the Air Force continued setting records with the lighter-than-air craft. On August 16, 1960, Captain Joseph Kittinger parachuted from an Excelsior gondola – a successor of the Man High craft – at an altitude of 102,800 feet. (US Air Force photo)

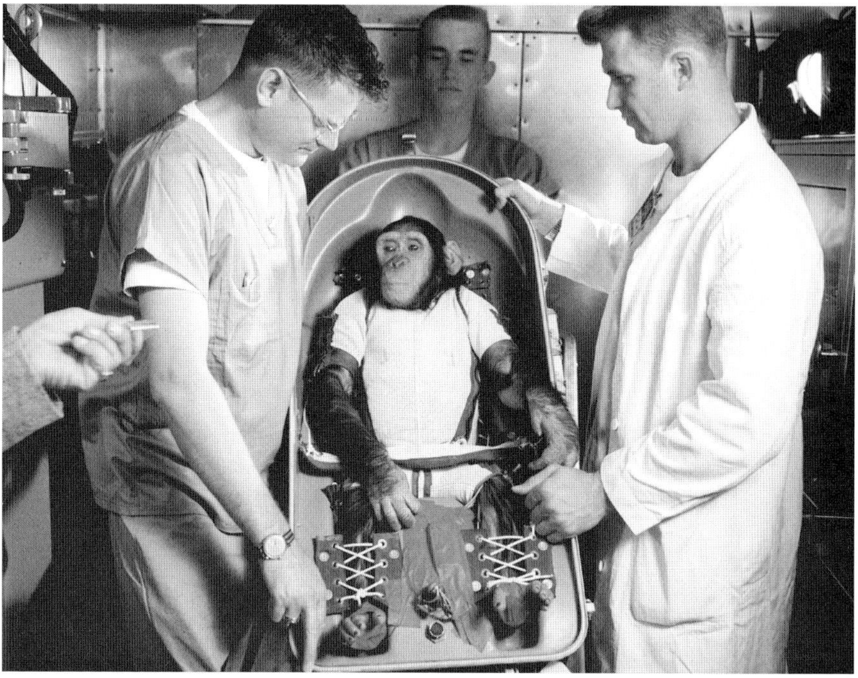

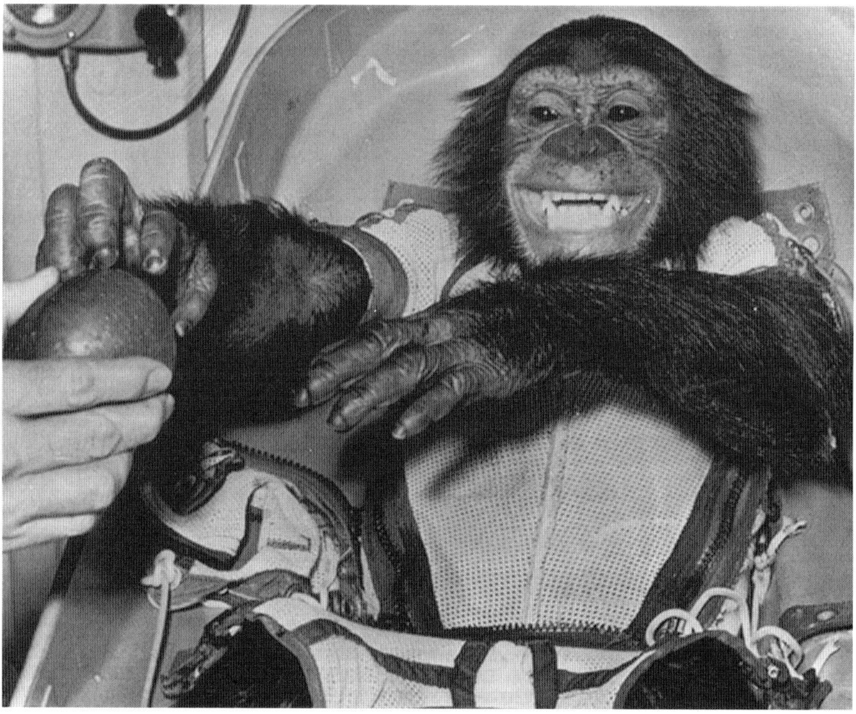

Opposite above: A four-year-old chimpanzee named Ham comprised the sole crew for the second Mercury-Redstone test on January 31, 1961. NASA trained Ham to pull functionless levers – roughly simulating the operation of a spacecraft – while staying calm in a small space undergoing radical changes in temperature and acceleration. Fail to pull a lever as required, and he got an electrical shock via his special crew couch. (NASA photo)

Opposite below: The force of acceleration on the second Mercury-Redstone mission on January 31, 1961 was supposed to peak at 12 Gs. But Ham the chimpanzee, the sole crew for the test-flight, actually endured nearly 15 times normal gravity. It also took three hours for US Navy crews to reach Ham after he splashed back down. By then, nearly half a ton of seawater was sloshing around inside his capsule. When the Navy finally hoisted Ham aboard the recovery ship, he happily gobbled an apple and half an orange. (US National Archives photo)

The second Mercury-Redstone mission on Jan. 31, 1961, pictured here, aimed to correct problems that occurred with the initial unmanned test of the capsule-rocket combination on December 19, 1960 – namely, the vehicle's trajectory resulted in a greater force of acceleration than the planners anticipated. Also, poor workmanship at the rocket shop was evident when washers, nuts and wire clippings freely floated inside the capsule during the period of weightlessness at the flight's apogee. In the end, the chimpanzee-crewed second Redstone test was cleaner, but still produced too much G-force. (NASA photo)

Chapter 19

I Aim at the Stars

In 1958 von Braun dialed up his decades-long effort to rewrite his own past. As his star rose in the mid-1950s, the German entertained offers to adapt his life's story into a movie which is where it got complicated. The US Army, complicit in von Braun's self-hagiography, reserved the right to vet the terms of any deal. A German film producer made the first offer – $24,000, that is $224,000 in 2021 dollars. Then Columbia Pictures in the United States bought its way into an increasingly complex international co-production at the cost of an additional $25,000 – $232,000 in 2021 – plus seven percent of profits.

The story treatment slowly took shape, each time running a gamut of producers, government censors and von Braun himself. 'As you know, they are anxiously trying to show that you were no Nazi, although you were a member of the Party and built the V-2 for Hitler,' one producer told von Braun.

I Aim at the Stars finally began shooting in late 1959. In the movie's version of von Braun's life, Hermann Oberth – a man von Braun admired but only briefly worked with – is a constant presence. Dornberger is there, too, as an adviser to the genius inventor toiling mostly alone on his rockets. Von Braun's wife Maria appears, but the producers aged her up by a decade. In reality, Maria was a girl during von Braun's time at Peenemünde.

Those deviations from the truth might be forgivable but they were incidental to the movie's main project, which was to depict von Braun as an unhappy slave of the Nazi war effort. 'From the beginning he is held up as a hero, a dedicated space pioneer, who is misunderstood by his parents, by his fiancée, by the German ordnance engineers and by the SS chief, Heinrich Himmler, who tries to make him out an enemy of the Third Reich,' *The New York Times* wrote in its review.

The producers anticipated this critique, as did the film's British director, J. Lee Thompson. 'Many Britons feel von Braun should have stood trial as a war criminal,' Thompson acknowledged.

Late revisions to the script added an American character, a US Army officer, whose family dies in a V-2 strike and who later becomes a journalist and critic of the fictionalized von Braun. The addition – and Thompson's own ambivalence – did little to dilute the film's oversweet portrayal of its subject.

American critics found the movie simplistic and boring. 'The balance of feeling for or against Dr. von Braun is not likely to be tipped in either direction by the peculiar emotionalism in this film,' the *Times* lamented.

In the United Kingdom, passions ran high. Protestors rushed the theater at the UK premier in London in December 1960. 'Nazi Braun's V-2 Rockets Killed and Maimed 9,000 Londoners,' the protestors' banner read. Stand-up comedian Mort Sahl added a satirical subtitle to von Braun's biopic: *I Aim at the Stars* 'but sometimes I hit London'.

I Aim at the Stars was a flop. Von Braun never got to collect his share of the profits, because there weren't any. In retrospect the movie's release perhaps represented the peak of von Braun's mainstream fame. From roughly May 1961, the month Alan Shepard became the first American in space, astronauts displaced rocket scientists – that is to say, von Braun – as the public face of the US space program.

President Kennedy helped to shift this paradigm. 'Kennedy was particularly drawn to the astronauts,' historian Logdson wrote. The Kennedys especially liked the handsome, tough, well-spoken

John Glenn. The astronaut was a frequent guest at the first family's compound in Hyannisport, Massachusetts. Glenn water-skied with Jacqueline Kennedy and talked the president into letting the astronauts sell their life stories to *Life* magazine.

Director Stanley Kubrick's 1964 film *Dr. Strangelove or: How I Learned to Stop Worrying and Love the Bomb* landed a more lasting blow. *Dr. Strangelove* is a satire in which a rogue US Air Force general sends a Communist-hating B-52 crew on a one-way mission to nuke the Soviet Union while the American president, his advisers and a fascistic, German-accented scientist – the eponymous Dr. Stangelove – plot the post-apocalypse.

Co-writer Terry Southern plainly told his biographer that he and Kubrick based the character of Strangelove on von Braun. In contrast to *I Aim at the Stars*, *Dr. Strangelove* was a hit. Five decades after its release, *I Aim at the Stars* is all but forgotten. Kubrick's movie, on the other hand, still is widely available.

Thus von Braun's star fell as those of the astronauts rose. Which is not to say von Braun was any less important to the space program after 1961. If anything, his importance grew as American efforts in space coalesced around a singular goal – to land men on the Moon, and do it before the Soviets did. But there was no denying that von Braun's story was changing. The ex-Nazi had labored for more than a decade to tell a tall-tale about himself. It was easier in the 1950s. His audience, a jingoistic American public, was forgiving. His editor, the US Army, helpfully reined in Von Braun's tendency toward occasional, accidental confession.

Hagiography got harder in the 1960s. The public was less motivated to buy a happy tale. Von Braun no longer worked for the Army so the Army no longer protected him. Von Braun bet big on *I Aim at the Stars* and it flopped, even backfired, if the protest in London is any indication.

Von Braun's real reckoning would come after his death but in the final 15 years of his life, he discovered the limits of his fame.

Alan Shepard's Freedom 7 mission represented a turning point for Wernher von Braun. As the first American in space, Shepard also was the first person in the US space program with more public appeal than von Braun had. Before 1961, von Braun was the face of the space program and its biggest advocate – this despite the German's Nazi past. After 1961, Americans had less morally-compromised space heroes to look up to. Von Braun's star fell as the astronauts' stars rose. In the 1950s, von Braun had starred in successful movies for Walt Disney. In the 1960s, he was the subject of a Hollywood flop. (Pictured here, Shepard's Mercury-Redstone rocket. NASA photo)

I Aim at the Stars, a fictionalized film portrayal of Wernher von Braun's life and work, represented the ultimate attempt to rewrite his history. In real life, he was an ambivalent, if not willing, participant in the Nazi regime's genocidal war effort. In the movie, von Braun essentially was a slave to the Nazis, one who constantly objected to the Nazis' aims and methods. Von Braun earned a small fortune selling his life rights to the film's American and German producers, but the movie was a flop for the studios. *I Aim at the Stars* aimed at hagiography, but missed. The tedious, unimaginative movie inspired a critical and public backlash, especially in the United Kingdom, where memories of von Braun's V-2 rockets – and their thousands of victims – were still fresh at the film's 1960 premiere. Pictured here, Curd Jurgens portrays a young Wernher von Braun at the site of a V-2 explosion in *I Aim at the Stars*. (Columbia Pictures capture)

Peter Sellers as Dr. Strangelove in Stanley Kubrick's classic Cold War satire *Dr. Strangelove or: How I Learned to Stop Worrying and Love the Bomb*. Co-writer Terry Southern plainly told his biographer he and Kubrick based the character of Strangelove on Wernher von Braun. (Columbia Pictures capture)

Chapter 20

Dyna-Soar

NASA's formation in 1958 and the gradual transfer of many military space programs to the new agency left intact, at least briefly, one major US Air Force initiative – an ambitious effort to develop an orbital outpost combining a hypersonic space-plane and a modular space station.

The space-plane and space-station combo, which never actually reached orbit in its full-up configuration, was the closest anyone came to realizing the most grandiose visions Oberth, von Braun and the husband-wife pair of Sänger and Bredt had described in the 1930s, 1940s and 1950s. It's no coincidence that the combo was a military effort. Oberth, von Braun and Sänger and Bredt sensed from the beginning that their conceptual space-planes and space stations lent themselves not just to war, but to *apocalyptic* war.

When the Air Force's space-plane and space-station project failed, so too did a possible version of history that could have ended very badly for the whole human race. If the Air Force's space-plane had cleared just a couple more bureaucratic hurdles, military astronauts might have been the very first Americans in space. If advocates of the flying branch's space station had convinced just a few more Pentagon budgeteers, the military would have occupied an enduring orbital habitat a whole decade before NASA finally managed to pull off the same feat. In other words, the Pentagon very nearly seized *de facto* control of the US space program despite Congress and successive presidential administrations deliberately choosing NASA, a civilian agency, to lead the United States into orbit.

It's impossible to say for sure how the Soviet Union would have responded to the US military seizing the initiative in the American side of the space race. But it's hard to imagine the Soviet response *reducing* tensions at a time, during the early-to-mid-1960s, when tensions hardly could have been higher without sparking global nuclear war.

The Air Force actually developed its space-plane and space-station under separate programs, but meant for them eventually to work together. The Dynamic Soaring space-plane, or 'Dyna-Soar', was a boost-glide system. It would *boost* into orbit atop a rocket then *glide* back to Earth like an unpowered airplane. The boost-glide concept combined the advantages of an ICBM, a satellite and a bomber.

A boost-glide space-plane could park in orbit for a hemispheric perspective on Earth and range far enough to drop atomic bombs anywhere on the planet on short notice, all while flying fast enough to dodge any conventional defense. It might even be reusable, a key quality for sustained wartime operations.

The Air Force formalized Dyna-Soar in 1957 and, in the six years before the Defense Department finally canceled the program, made a perhaps surprising amount of progress toward an actual, functional vehicle. When a boost-glide space-plane belatedly reappeared in the United States a decade and a half later, it was in the form of an arguably much less elegant NASA system – the bulky, fragile Space Shuttle.

A boost-glide space-plane might seem futuristic by 1950s standards, but in fact the idea pre-dated Dyna-Soar by many decades. Von Braun had tinkered with the concept by way of the winged V-2, which he hoped might evolve into a manned space-plane. But boost-glide was older even than that. The first major boost-glide proposal came from the Austrian husband-wife team of Eugen Sänger and Irene Bredt. In 1933, Sänger and Bredt sketched a notional, rocket-propelled 'antipodal bomber' that would launch horizontally from a sled, accelerate to 13 times the speed of sound while climbing through the Karman Line.

Slowing and plunging back toward Earth, the 111-foot-long bomber would bounce off the upper atmosphere like a Mach-three stone skipping on water. It would drop its one-ton bomb on a target as far as 15,000 miles away then return for a gliding landing at the same base where it had launched.

Sänger and Bredt got as far as building a scale model for wind-tunnel tests before the Third Reich balked. The Nazis rarely met a high-concept weapons program they didn't love, but the Sänger-Bredt antipodal bomber struck even them as far-fetched.

Two key figures were more impressed. One was Theodore von Karman, a Hungarian-born scientific advisor to US Army Air Force general Hap Arnold. If von Karman's name sounds familiar, it's because he was the first to calculate the altitude where the atmosphere ended and space began: the Karman Line. At the war's end, von Karman led a team on a fact-finding tour of Europe on behalf of Arnold and the Army Air Force. Von Karman toured German aeronautical facilities, interviewed von Braun and Dornberger and gathered up lots and lots of documents.

He came across a study from Sänger and Bredt describing their antipodal bomber. Von Karman folded Sänger and Bredt's work into his reports to the Army and, later, the Air Force. He apparently also had the Sänger-Bredt concept in mind when, in 1948, he helped to found the Research and Development Corporation think tank, or RAND, in California. In May 1946, RAND handed over to the Army a key study on the future of US military aeronautics. The study strongly endorsed military satellites as observation platforms and bombers, but didn't specify exactly how those satellites should reach orbit.

Von Karman and RAND's work laid the foundation for a boost-glide space-plane in the United States. Meanwhile, Soviet scientists were laying a similar foundation on the opposite side of what would become known as the Iron Curtain, for the *other* party besides von Karman to take a strong interest in Sänger and Bredt's work was Soviet premier Joseph Stalin. The Soviet strongman was so enamored of the hypothetical space-plane that he dispatched a team to western Europe to locate Sänger and Bredt and either convince them to work for the Soviets – or kidnap them.

Stalin's team failed. But the dictator's interest in the boost-glide vehicle endured. He tasked Soviet scientists with developing a space-plane roughly matching the Sänger-Bredt antipodal bomber. That work continued until 1953, when Korolev – the rocket-scientist Stalin had imprisoned then freed – completed work on the R-7, the USSR's first ICBM and the basis of the space-launch-vehicle that would deliver the Sputnik satellites into orbit. Stalin was satisfied that he had a working launch vehicle and strategic weapon with global range so with R-7 in hand, he didn't need an antipodal bomber. Work on a boost-glide system called 'Zvezda' effectively ended in the late 1950s when Soviet authorities finally released and returned to East Germany the last of the former Peenemünders they had recruited in competition with the American Operation Paperclip back in 1945.

The Americans stuck with it, though. Ironically, bureaucratic confusion and competition in the aftermath of NASA's 1958 founding likely helped to sustain US boost-glide efforts. As NASA absorbed more and more of the country's space projects, the Air Force clung more tightly to whatever programs it still had in its column. That included the boost-glide vehicle. While von Karman and RAND, borrowing heavily from Sänger and Bredt, helped to define what might be possible with a boost-glide space-plane, it was Dornberger who gave the concept shape.

After the British government finally released him in 1947, Dornberger followed von Braun to the United States. He spent three years advising the Air Force on missile-guidance then, in 1950, took a plum job with the Bell Aircraft Corporation. Bell's most famous product was X-1, the first aircraft to fly faster than the speed of sound.

Dornberger provided to Bell a detailed concept for a boost-glide bomber and in 1952, Bell took its bomber-missile, or BOMI, to the

Air Force. BOMI was the antipodal bomber in everything but name. It would accelerate to a peak velocity of Mach 12 before settling into a Mach-four cruise over a range of up to 5,000 miles. Its payload would be a 7,000-pound nuclear bomb.

Air Force leaders were aware of Stalin's interest in a boost-glide space-plane and were keen to at least keep pace with the Soviets' own hypersonic efforts. The Air Force imagined it could develop and field BOMI by 1962. Incredibly, it got close.

RAND gazed into the proverbial crystal ball and concluded that rocket technology would soon advance to the point where nuclear-tipped ICBMs could take over from manned bombers as the United States' main nuclear deterrent. The think-tank worried that ICBMs would lack accuracy, reliability and yield. America would continue to require manned bombers for hunting down, and nuking, the most elusive Soviet forces.

But increasingly sophisticated air-defenses posed an existential threat to the traditional, air-breathing manned bomber. A space-plane however could fly over – and outrun – any conceivable surface-to-air missile. In addition to flying high and fast, an antipodal bomber was maneuverable – and that made it unpredictable. That unpredictability gave the boost-glide vehicle enduring value, even as ICBMs proliferated.

Early ICBMs lacked true global range. To reach the Soviet Union from the United States or vice versa, they traveled over the North Pole and this made them predictable. Even early submarine-launched ballistic nukes were fairly easy to anticipate, as the subs needed to close within a thousand or miles or so of their targets. This narrowed their options. The first-generation early-warning systems that the United States and Soviet Union developed in the early 1960s had gaps. The American system didn't have much in the way of south-gazing sensors. The Soviet system mostly looked north and, to a lesser extent, south. It was blind to the east and west. An antipodal bomber could exploit these gaps.

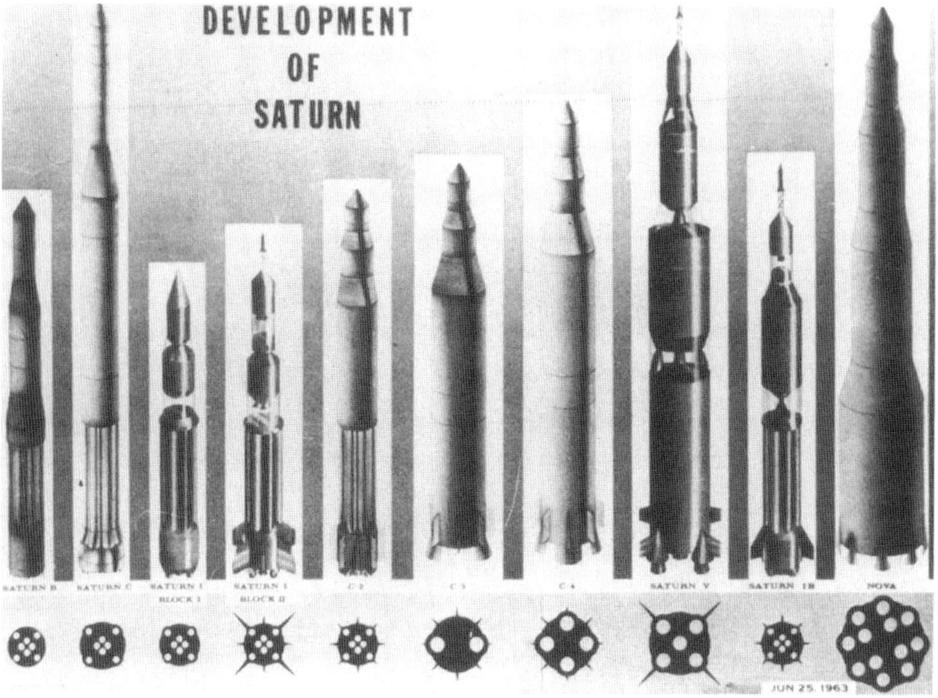

Above: Saturn wasn't one rocket, it was a family of rockets, each more powerful than the one before it. The ultimate planned version of Saturn was the C-8, which was more or less identical to the Nova, a three-stage booster generating an incredible 16 million pounds of thrust – enough, NASA figured, to fling a spacecraft all the way to Mars. NASA abandoned that super-huge booster around 1964 as it became clear that the Moon was as far as manned missions were going to travel, at least for a few generations. In the end, NASA built just three versions of Saturn of increasing size – the two-stage Saturn I and two separate three-stage models, Saturn IB and Saturn C-5. The latter is better known by its other name, Saturn V. (NASA art)

Opposite above: Eugen Sänger and Irene Bredt were decades ahead of the curve when, in 1933, they proposed the first boost-glide space-plane, their so-called 'antipodal bomber'. Sänger and Bredt got as far as building a scale model for wind-tunnel tests before the Third Reich balked. Theodore von Karman, a Hungarian-born scientific advisor to the US Army Air Force, was more impressed. He helped to popularize the concept in the United States. The Sänger-Bredt antipodal bomber inspired generations of boost-glide space-planes, most notably the US Air Force's Dyna-Soar space-plane. (Eugen Sänger and Irene Bredt art)

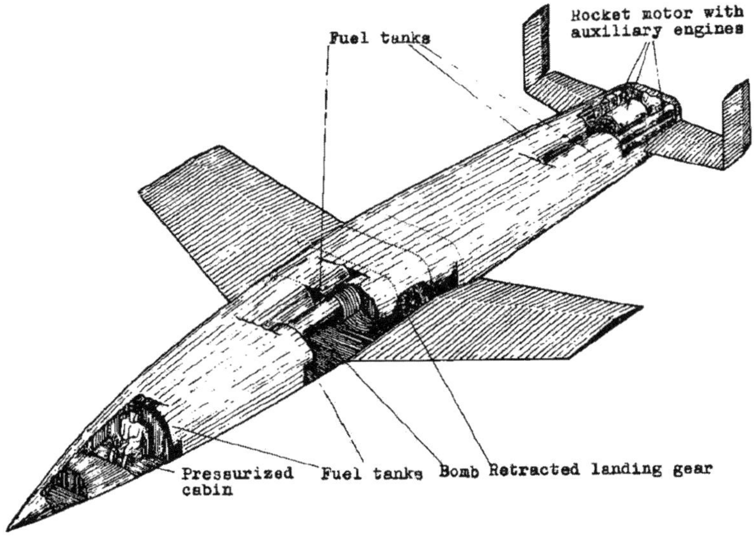

Figure 33; Total view of 10 ton Rocker Bomber

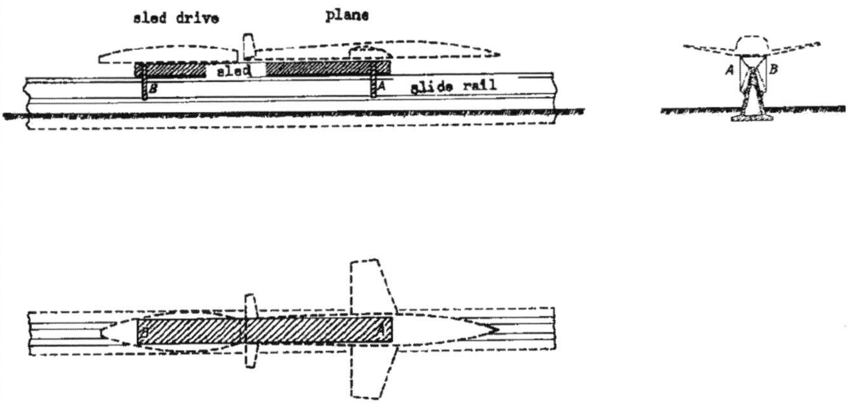

Eugen Sänger and Irene Bredt's boost-glide antipodal bomber launched via a sled. It accelerated to 13 times the speed of sound while climbing through the Karman Line separating the atmosphere from space. Slowing and plunging back toward Earth, the 111-foot-long bomber would bounce off the upper atmosphere at a cruising speed of Mach 3. It would drop its one-ton atomic bomb on a target as far as 15,000 miles away then return for a gliding landing at the same base where it launched. (Eugen Sänger and Irene Bredt art)

There was at least one major obstacle. Boosting a space-plane to the Karman Line demanded a more powerful rocket than the United States possessed in the mid-1950s. BOMI wasn't alone in this regard. A lot of American space projects required more lifting power than the first generation of launch vehicles offered. As the Air Force studied different configurations for its atmosphere-skipping bomber, it briefly considered von Braun's Saturn – then little more than an idea – as the booster.

Wernher von Braun with a model of a Jupiter rocket. Von Braun proposed strapping eight Redstones to a central Jupiter in order to create a so-called 'Super Jupiter' that would be big enough to meet America's space-exploration needs through the 1960s. Instead, the Pentagon assigned von Braun to develop Saturn, the ultimate version of which was even more powerful than Super-Jupiter would have been. (NASA photo)

Chapter 21

'We choose to go to the Moon'

The Air Force didn't just hand the space-plane contract to Bell. The service had options. No fewer than four separate boost-glide program offices and a dozen aerospace firms were working in parallel on boost-glide designs. Finally in 1957, the Air Force collapsed all the competing efforts into one program, Dyna-Soar. The goal was, by 1974, to produce a boost-glide vehicle with a maximum speed of Mach 20, a ceiling of 300,000 feet and a range of 10,000 miles.

Rather than skipping off the atmosphere, per Sänger and Bredt's proposal, Dyna-Soar would simply boost through the Karman Line … then glide back down in a simple, albeit fiery, arc.

The bomber-missile would be capable of several missions – nuclear bombardment, reconnaissance and 'satellite-inspection'. The latter was something of a misnomer; not only should Dyna-Soar be able to survey an enemy satellite, it should be able to disable it, as well.

The competition to develop the Dyna-Soar came down to two companies. Bell, of course – which owing in part to Dornberger's expertise already had a pretty good idea what an operational space-plane should look like as the New York firm started pitching space-plane concepts to the Air Force as early as 1952.

Then there was Boeing. The Chicago plane-maker came late to Dyna-Soar, but the design it proposed for the Dyna-Soar program had the advantage of being lighter than its rivals. That in turn meant less demand on the rocket that would accelerate the space-plane to hypersonic speed. In June 1959, the Air Force tapped Boeing to build

Dyna-Soar. Considering how advanced Bell's space-plane design was compared to its competitors' own designs, it was a controversial decision. Indeed, as Boeing's blueprints for the operational Dyna-Soar evolved between 1959 and the design freeze in early 1960, they started looking more and more like Bell's own blueprints.

Dyna-Soar took shape during the post-Sputnik crisis in US space policy. President Eisenhower bowed to pressure from the public and Congress to take strong action to boost America's efforts in space. 'After Eisenhower failed to show the leadership necessary to manage the Sputnik situation, he found pressure from a set of political exigencies, a critical mass of interests, and a key cadre of scientific and technical officials within the federal government and their supporters in universities, corporations and think-tanks to create a powerful, large and costly federal agency to carry out a more aggressive space exploration program,' historian Launius explained. 'In so doing, Eisenhower was manipulated by these forces to establish a separate federal entity, something he thought unnecessarily expensive and once created almost impossible to dismantle.'

The president reluctantly sided with his critics on a proposal, soon codified in law, to fold the country's scattered, bickering space programs – currently scattered across the armed services – into one coherent, national effort under a *civilian* agency. For all Eisenhower's doubts, it turned out to be the most important decision in the whole space race. The president was right, once Congress acted to establish a space agency specifically in response to Sputnik, the agency would prove impossible to dismantle even after the panic over Sputnik had subsided.

That is to say, NASA survived the small-government Eisenhower administration. So it was in a position to expand fast and act quickly under incoming big-government President John F. Kennedy and his pro-space vice President Johnson, the same Johnson who, as a senator, had made political hay attacking Eisenhower for botching the US response to Sputnik.

Kennedy entered the White House in January 1961 determined to land Americans on the Moon. Eight days after Gagarin orbited Earth on April 12, Kennedy sent Johnson to meet with Von Braun. Johnson asked how fast Americans could get to the Moon and whether they could beat the Soviets.

'We have a sporting chance,' von Braun responded in writing. 'With an all-out crash program I think we could accomplish this objective in 1967 to 1968.' Kennedy and Johnson and their allies in Congress moved fast to fund that 'all-out crash program'.

In the meantime, however, Kennedy tried a hail-Mary play. 'Why don't we do it together?' he asked Krushchev at a summit in Vienna the summer following Kennedy's inauguration. Krushchev eventually demurred. A cooperative Moon mission was unrealistic as long as the USA and USSR were pointing tens of thousands of nuclear weapons at each other, the Soviet premier said.

Kennedy's wife Jacqueline chatted up the recalcitrant Khrushchev at a state dinner. She ended up asking if she could have one of the puppies of a Soviet space dog. Jackie wasn't serious. 'I was just running out of things to say,' she recalled. Krushchev didn't know that. A couple months later, the Soviet ambassador to the United States showed up at the White House with a 'terrified small dog', according to historian John Logsdon.

The Kennedy-era burst of political capital, as well as literal capital in the form of billions of dollars in new funding, boosted NASA at a critical moment when US military space programs were struggling. Just as Eisenhower feared, NASA steadily expanded through the 1960s, its budget swelling from $89 million in 1958 ($800 million in 2021 dollars) to a staggering $5.9 billion ($42 billion in 2021 dollars) in 1966. Counting contractors, in 1966 NASA employed nearly 400,000 people.

Even a catastrophic accident – a fire aboard the Apollo 1 capsule that killed astronauts Gus Grissom, Ed White and Roger Chaffee on the ground at Cape Canaveral on January 27, 1967 – did little to slow NASA's momentum.

All that was bad news for tax-cutting right-wingers, but good news for the solidifying majority of Americans, and American politicians, who prioritized at least keeping up with the Soviets in space, if not beating them. It also was good news for advocates of the peaceful use of space. By the early 1960s it was NASA, rather than the Army, Air Force or Navy, leading America into space. But NASA didn't stand up overnight. In the short term, it added yet another bureaucracy to the already-fragmented US space establishment. The Air Force's Mach-20 bomber first benefited from that confusion before ultimately falling victim to it. It was a military project at a time when civilians were assuming control over US space exploration. Moreover, it was a manned, armed vehicle that – while notionally suborbital – clearly would be capable of reaching orbit.

If Dyna-Soar succeeded, and on schedule, it might beat NASA's capsule into space. The White House, for all its ambivalence toward NASA, was opposed to sending the first American into orbit aboard a *bomber*. The Air Force wisely pledged to keep Dyna-Soar below the Karman Line during early testing, thus affording NASA a chance to get to space first.

Kennedy's inauguration in January 1961 signaled even greater tumult for Dyna-Soar. While the Kennedy administration by and large continued the space efforts of the outgoing Eisenhower administration in order to expand on them, Kennedy's defense secretary Robert McNamara introduced an impatient dynamic to defense and space planning. McNamara famously came from a business background. His team of efficiency-minded 'whiz kids' demanded swift responses to blunt questions and took pride in making snap decisions based on what they insisted was clinical analysis. Decades later McNamara, like his distant successor Donald Rumsfeld, has a reputation for unsound policymaking.

In fact, McNamara's ruthless efficiency-seeking had the effect of blocking the Air Force's best effort to undercut NASA and contest control of the US space program. Paradoxically, however, the Air Force space-plane project actually represented a temporary lifeline for von Braun's Saturn super-booster, which later would prove critical to NASA's successes in orbit and beyond. Von Braun started drawing up blueprints for Saturn in the late 1950s while he was still with the Army. The service sensed it, and the country would eventually need a superheavy booster, but in the early years of Saturn's development, neither the ground-combat branch nor one of its sister services could identify major applications for the giant rocket.

The Dyna-Soar was one of the few payloads that might require the millions of pounds of thrust Saturn could generate. ARPA, forever trying to mediate between the squabbling armed services, leaned on the Air Force to use Saturn with Dyna-Soar. By January 1961, von Braun had completed an initial study of the Saturn and Dyna-Soar combo and was ready to cut contracts and actually build the giant booster.

But things were changing and fast. The plan was to test Dyna-Soar in the atmosphere before boosting it over the Karman Line but Boeing for one seemed to appreciate just how long that testing might take – and how impatient McNamara's Pentagon might become in the meantime. To please McNamara and the whiz kids, the company drew up a new testing plan that skipped the suborbital flights.

There were even bigger problems with Dyna-Soar, namely heat. A review of the space-plane's design indicated the crew compartment would get too hot for delicate electronics – to say nothing of a human crew. Boeing had to redesign Dyna-Soar with new ceramic material, special insulation and high-tech cooling systems. McNamara still wasn't convinced. If NASA was going to lead America into orbit, why did the Air Force need a space-plane? Especially when that space-plane's development was such a mess.

It didn't help that in August 1960, the Eisenhower administration despite its government-shrinking philosophy had created a secretive organization, the National Reconnaissance Office, to oversee the country's new Corona spy satellites, the first of which launched in

June 1959. Historian Logsdon theorized that Eisenhower consented to stand up NASA in part because he thought the agency would draw attention away from important military space projects. In as much as the NRO remained an official secret until 1992, the gambit may have worked – perhaps too well. In the early 1960s, even the Air Force didn't know exactly what the NRO was up to.

That secrecy complicated the flying branch's efforts to sell the White House on the Dyna-Soar program. Air Force leaders couldn't say for sure whether the NRO already possessed, or soon would possess, the same capabilities the service was hoping Dyna-Soar might offer.

The Soviet Union inadvertently dealt Dyna-Soar another blow when, on April 12, 1961, a Vostok rocket – a version of the liquid-oxygen-and-kerosene-fueled R-7 – boosted cosmonaut Yuri Gagarin into orbit for 108 minutes, the first human being in space. Like the Sputniks before it, the Vostok 1 mission was less impressive than it seemed. OKB-1's engineers hadn't figured out how to safely land a re-entry vehicle with its occupants still aboard. So in classic Soviet fashion, they developed a simple workaround. At 20,000 feet, Gagarin bailed out of the heat-scorched re-entry vehicle and parachuted to a safe landing near a farmer's field in southern Russia. But even that work-around nearly failed when Gagarin's capsule tumbled and heated up faster than its designers had anticipated.

Vostok 1 beat NASA's own first man-in-space mission – the suborbital Mercury-Redstone 3, boosting astronaut Alan Shepard in the Freedom 7 capsule by a mere month. In stark contrast to the Soviet re-entry vehicle, Freedom 7 safely could bring its occupant all the way to the surface. Still, Gagarin's feat had immense propaganda value. The Americans didn't boost their own astronaut into Earth's orbit until February 20, 1962, when John Glenn in the Friendship 7 capsule rode a Mercury-Atlas rocket past the Karman Line.

The Kennedy administration had just suffered its first humiliation in space. It was determined never to lose to the Soviets ever again. A little over a year after Vostok 1, on September 12, 1962, Kennedy would stand before a crowd in Houston, Texas and utter the unforgettable words, 'We choose to go to the Moon'.

Realistically, only von Braun's Saturn could get them there. The German's super-rocket went from being a technology in search of a mission to one with clear and urgent purpose. The Air Force was about to lose access to Saturn. The giant rocket had more important things to do than lob an experimental military space-plane into orbit on trials the Pentagon itself couldn't entirely endorse. But that didn't stop the flying branch from doubling down on Dyna-Soar. It was one thing to boost a man into orbit for an hour or two then have him parachute back down, Air Force officials reasoned. A true spacefaring nation would fly men into orbit as a matter of routine aboard reliable, reusable, multi-mission space-planes.

It's not that McNamara disagreed, *per se*. He wanted a military spacecraft to be capable of maneuvering in orbit in order to rendezvous with enemy craft before precisely landing. The same vehicle should be capable of another mission after a quick refurbishment. McNamara also wanted the Pentagon to work up to an enduring presence in space for testing and observation. If Dyna-Soar could fulfill some or all of those requirements – great! But if some other program or programs could meet those needs more quickly and at lower cost, all the better.

In February 1962, McNamara endorsed a plan to launch the first manned Dyna-Soar flight in late 1965 at a cost of $700 million – that's $6 billion in 2021. He also wanted the Air Force to work on an orbital station. In August that year, the Air Force cut a contract with Douglas Aircraft Company and General Electric to begin outlining a concept for a so-called 'Manned Orbital Laboratory'. The plan was for these two efforts eventually to merge. An evolved version of the Dyna-Soar with a passenger compartment for four people would fly back-and-forth on resupply missions for the MOL space station in low orbit.

At the same time, and separately, NASA – specifically, the agency's Langley Research Center – was drawing up detailed plans for its own space station. The 75-foot-diameter, 85-ton hexagonal station combined a rigid frame with inflatable habitats. Langley assumed it would pack the station into the nose cone of a Saturn rocket for the trip to low Earth orbit. Once deployed, the station would require regular visits. For this mission, NASA even briefly considered adopting a variant of Dyna-Soar, a space-plane that could work as a station support craft. That fleeting proposal actually signaled the end of NASA's interest in Dyna-Soar.

The Air Force initiated the Dyna-Soar program in 1957 and canceled it six years later owing to shifting priorities and NASA's growing monopoly on manned space missions. When boost-glide reappeared in the United States, it was in the form of a NASA program. The Space Shuttle. A mock-up of the Dyna-Soar's cockpit. (US Air Force photo)

On April 12, 1961, a Vostok rocket – a version of the R-7, the USSR's first ICBM – boosted cosmonaut Yuri Gagarin into orbit, making him the first human being in space and marking yet another victory for the Soviet space program over the rival American program. Depicted here, the business end of an R-7. (RKK Energia photo)

Alan Shepard in the Freedom 7 capsule before his historic launch on May 5, 1961. After a decade heading NASA's astronaut training, Shepard returned to space in 1971 as the commander of Apollo 14, NASA's third Moon-landing mission. Shepard was 47 years old when he finally stepped foot on the Moon, making him the oldest lunar explorer. (NASA photo)

Above: A Mercury-Redstone rocket boosted astronaut Alan Shepard to an altitude just below the Karman Line on May 5, 1961. The Soviets had gotten there first, and further. On April 12, 1961, a Vostok rocket had carried cosmonaut Yuri Gagarin *past* the Karman Line. (NASA photo)

Opposite above: This painting by cosmonauts Alexei Leonov and Andrei Sokolev depicts Yuri Gagarin parachuting to Earth after bailing out of the Vostok 1 re-entry vehicle. (Art via NASA)

Wernher von Braun rides in the back of a motorized cart during President John F. Kennedy's tour of the McDonnell Aircraft Corporation plant in St. Louis, Missouri. Kennedy visited the plant on September 12, 1962 as part of a two-day inspection of NASA facilities. The same day, Kennedy stood before a crowd in Houston and uttered the unforgettable words, 'We choose to go to the Moon'. (US National Archives photo)

Above: For three years starting in 1958, Wernher von Braun worked on the design for an initial version of the Saturn heavy rocket. The US Air Force's Dyna-Soar space-plane was the main payload. By January 1961, von Braun had completed an initial study of the Saturn and Dyna-Soar combo and was ready to begin actually building test models of the giant booster. Von Braun is depicted here in conversation with an unidentified man during President John F. Kennedy's visit to Cape Canaveral on September 11, 1962. (US National Archives photo)

Opposite above: A single MOL – the Air Force envisioned a whole constellation of them – was little more than a 60-foot-long tube that was the same diameter as a Gemini capsule. This cutaway is from a 1967 planning document. (US Air Force art)

Opposite below: President John F. Kennedy meets with his controversial secretary of defense Robert McNamara at the White House in June 1962. McNamara's insistence on efficiency often backfired and instead sowed confusion inside promising technology programs such as the Dyna-Soar space-plane. (John F. Kennedy Library and Museum photo by Cecil Stoughton)

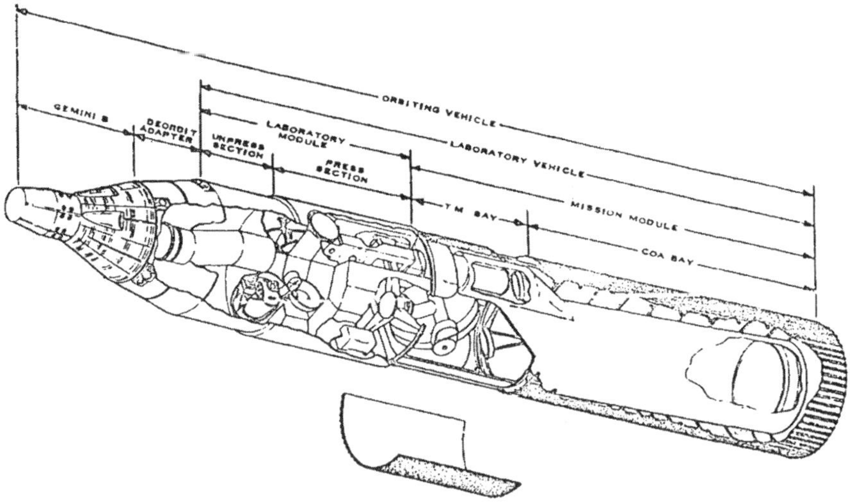
GEMINI B
DEORBIT ADAPTER
UNPRESS SECTION
LABORATORY MODULE
PRESS SECTION
ORBITING VEHICLE
LABORATORY VEHICLE
MISSION MODULE
TM BAY
COA BAY

Lives were at stake in the late 1950s and early 1960s as Wernher von Braun designed, built, tested and refined the giant Saturn rocket. Early on, Saturn was destined to boost the US Air Force's Dyna-Soar space-plane, but Dyna-Soar fell victim to Pentagon cost-cutting, leaving Saturn with one main payload – Moon-bound astronauts. In September 1962, von Braun – by then the director of the Marshall Space Flight Center in Huntsville, Alabama – invited five astronauts and Brainerd Holmes, a NASA associate administrator, to a meeting in Huntsville. From the left are astronauts Elliot See, Tom Stafford, Wally Schirra and John Glenn, Holmes, von Braun and astronaut Jim Lovell. (NASA photo)

Chapter 22

Von Braun's mega rocket

In 1961 and 1962 the Air Force and NASA were talking past each other, each focusing on one half of a station and space-plane combo but, of course, failing to fully cooperate. Why would they? The law and political momentum were behind NASA. Hazy dreams of NASA Dyna-Soars supplying some future, inflatable space station aside, the space agency didn't need very much from the Air Force in order to remain in charge of America's space program. The Air Force was fighting a rearguard action to hold on to any meaningful role in space.

The main thing the Air Force and NASA shared in the early 1960s was von Braun's Saturn rocket. Through 1961, there were different Saturn configurations for different missions. The C-1 configuration originally had three stages, then it had two. There was a three-stage C-2 configuration until there wasn't. A C-3 configuration, also with three stages, replaced the C-2. There was a C-4 and a C-5, both superheavy, three-stage configurations. Dyna-Soar would have launched atop the C-3 version of Saturn.

By late 1962, Kennedy's Moon mission was the big thing. Marginal space efforts faded as the Moon push gobbled up manpower and cash. NASA eventually cut back on Saturn configurations and also walked away from a second heavy rocket the agency called 'Nova'. If a program didn't help Americans get to the Moon, it tumbled down the list of priorities. Langley's inflatable space station was one victim. The Saturn C-3 was another. Its elimination did nothing to help Dyna-Soar. From NASA's perspective, the Air Force space-plane 'seemed to evaporate'. In the end, there were three models of Saturn, each more powerful than the previous. The two-stage Saturn I, the three-stage Saturn IB and the ultimate Saturn V, also with three stages.

After 1962, NASA no longer considered Dyna-Soar a Saturn payload. Still, the space-plane project survived through 1963. Indeed, it's not clear the Air Force really minded losing Saturn. Getting cut off from von Braun's super-booster justified the flying branch developing its own new heavy rocket, the Titan III, which it could build exclusively for military missions. Titan III was a version of the Titan I ICBM but with strap-on boosters for extra thrust. It was a hybrid launcher that burned both solid and liquid fuels, and with its heavy payload performance it threw Dyna-Soar a lifeline – albeit a short one.

Dyna-Soar was adding cost and distraction to US space efforts at a time when the twin forces of McNamara and the Moon push demanded efficiency and focus. By June 1963 McNamara had made up his mind. The Air Force didn't need Dyna-Soar, not when it could just copy NASA's Gemini spacecraft, paint it blue and send it on military missions.

McNamara canceled Dyna-Soar in December 1963. 'Blue Gemini' should replace Dyna-Soar for quick trips to orbit, the defense secretary insisted. The new Manned Orbital Laboratory – a 42-foot-long, 10-foot-wide space station that would travel into orbit underneath a Gemini capsule – would assume Dyna-Soar's experimental duties and also function as a high-endurance observation platform.

The cancelation brought to an end nearly a decade of tumult and confusion as the Air Force fought for a toehold in space. Tragically for the thousands of people who worked on Dyna-Soar, McNamara killed off the program just as its major components were coming together. In that sense, Dyna-Soar was one of several programmatic dead-ends that steadily choked the Air Force's space program at a critical time when NASA, bolstered by von Braun's transfer from the Army, was growing in size and influence.

Dyna-Soar could have been. Boeing had solved the space-plane's heat problem, assembly of the first vehicle was underway. Dyna-Soar might have been ready to fly as early as October 1964. Instead, work stopped abruptly and the Air Force and its contractors shifted people and money into Blue Gemini and the new space station.

While the Air Force took pains to characterize the Manned Orbital Laboratory as a research station, in fact the service – and, perhaps more importantly, the NRO – was most interested in the station's potential as a reconnaissance platform. By 1962 the NRO had quietly deployed no fewer than 25 Corona spy satellites. More and better satellites were in development for the NRO and the Air Force, but as fast as satellite technology was improving there remained some seemingly intractable limitations. Namely, satellites moved relative to Earth's surface along fairly fixed and highly predictable orbits. It wasn't possible to reposition a satellite, at least not quickly. Early satellites sometimes struggled to focus their cameras. Weather was another persistent problem, all it took to obscure a target was a few clouds.

After a year of study, the Air Force, the NRO and McNamara's office concluded that a manned recon platform held certain advantages over an unmanned spy satellite. Harold Brown, McNamara's head of research and development, underscored a crew's 'ability to recognize patterns and interpret them in real-time and report the results' as well as 'the ability to point a sensor and provide image-motion compensation so that a very narrow field of view can encompass the area to be examined'.

It might have been true, in 1963, that an astronaut in a space station, pointing a camera, could shoot better photos than a robot could. But would that still be true after the years it would take to develop, build and launch a space station? No one knew. So work on the Manned Orbital Laboratory got underway.

The design came together quickly. A single MOL – the Air Force envisioned a whole constellation of them – was little more than a 60-foot-long tube that was the same diameter as a Gemini capsule. A Titan III would boost, as a single payload, a MOL and an attached Gemini, with the two station crew riding in the capsule and wearing special, extra-flexible suits that allowed them to squeeze through the hatch between station and capsule.

The Air Force designed MOL to support the two-man crew for 30 days. The station was, by the standards of later space stations, quite roomy. Each occupant had around 400 cubic feet of space. Beds were at one end of the lab, storage was at the other. A kitchen and toilet occupied one wall. The other held the workstations and a powerful Dorian camera with a 72-inch lens. A passageway at one end connected to the capsule that shuttled the crew to and from the station.

By November 1966, the Air Force had something – although not much – to show for its work. On November 3, a Titan III blasted off from the recently renamed John F. Kennedy Space Center at Cape Canaveral carrying an incomplete MOL mockup attached to a refurbished Gemini capsule that, way back in January '65, had flown an unmanned suborbital mission.

The MOL mockup lacked a KH-10 camera but did carry three satellites and a bevy of small science experiments. The crewless station safely separated and achieved orbit. The satellites launched without incident. The experiments ran their course.

The Air Force had hoped the half-complete MOL might last 75 days in orbit but it only lasted 30 before its radios failed. The station's orbit decayed and, in January 1967, it burned up in Earth's atmosphere.

Unbeknownst to the Air Force, the anticlimactic test was the end of something more than it was a beginning. Satellite technology was getting a lot better, a lot faster than anyone had anticipated when the Pentagon, the NRO and the Air Force endorsed the military space station.

MOL was losing relevance. It didn't help that the program was over-budget and behind schedule. In 1969, the incoming Nixon administration was looking for bloat in the federal budget. It found MOL, the biggest piece of hardware in a military space program that was growing more redundant by the day as NASA went from strength to strength. Nixon canceled MOL in June 1969. A month later, NASA landed astronauts on the Moon.

Dyna-Soar had fallen victim to McNamara's ruthless pursuit of efficiency. The 'efficient' Gemini-supported space station that replaced it died of irrelevance when it took too long to complete. The twin failures of Dyna-Soar and MOL doomed the Air Force's faltering effort to gain a toehold in manned spaceflight.

To be clear, the military was in space to stay. But the only craft it would own would be unmanned satellites. When military personnel traveled into space, it would be aboard NASA vehicles as part of NASA crews. In as much as the space race was a three-way contest between the United States and the Soviet Union and between America's civil and military agencies, NASA had won. The agency's victory significantly slowed the militarization of space, dealing a seatback to a notion of space exploration that had its long roots in the blood-soaked underground labs and slave camps of Nazi Germany. There would be no atmosphere-skipping bombers, no military outposts on Earth orbit.

And it should go without saying that exploding a nuclear warhead on the Moon or any similar show of force in space was off the table.

For his part, von Braun never had cared *how* his adopted country conquered the heavens. When Germany was his home, he happily hitched his extraterrestrial wagon to the Nazi regime's war horse. After he became an American, he rode the Pentagon's space program until NASA began winning the bureaucratic battles. After 1960, von Braun was a NASA man. He happily remained a NASA man until the money ran out 12 years later.

Above: President John F. Kennedy and Vice President Lyndon Johnson with Wernher von Braun during a tour of Redstone Arsenal in Huntsville, Alabama on September 11, 1962. A Saturn C-1 rocket is in the background. There were different Saturn configurations for different missions. The C-1 configuration originally had three stages, then two. (US National Archives photo)

Opposite above: US aerospace firms Bell and Boeing closely competed for the US Air Force contract to develop the Dyna-Soar space-plane. Bell Aerospace had more experience, but Boeing won anyway – perhaps because its design was the lightest and thus potentially the smallest burden for the rocket that would boost it. Boeing's March 1958 configuration, depicted here, changed a lot over the next two years. (Boeing photo)

Opposite below: By December 1958, Boeing had refined its Dyna-Soar design to reflect the extreme temperature and stress the space-plane would have to endure during re-entry. That required either a very strong vertical tail or smaller wingtip stabilizers. Boeing opted for the latter. (Boeing photo)

USAF

U.S. AIR FORCE
DOWN RANGE COMPLEX

By April 1959, Boeing had further refined the Dyna-Soar, bringing the space-plane close to its final configuration before its abrupt cancellation – some would say 'premature' – in December 1963. Irritatingly for executives and designers at Bell Aerospace, Boeing's final Dyna-Soar design looked a lot like Bell's much earlier configuration for the space-plane. (Boeing photo)

US Senator George Smathers of Florida and President John F. Kennedy receive a briefing on the Saturn rocket at a launch pad at Cape Canaveral on November 16, 1963. (US National Archives photo)

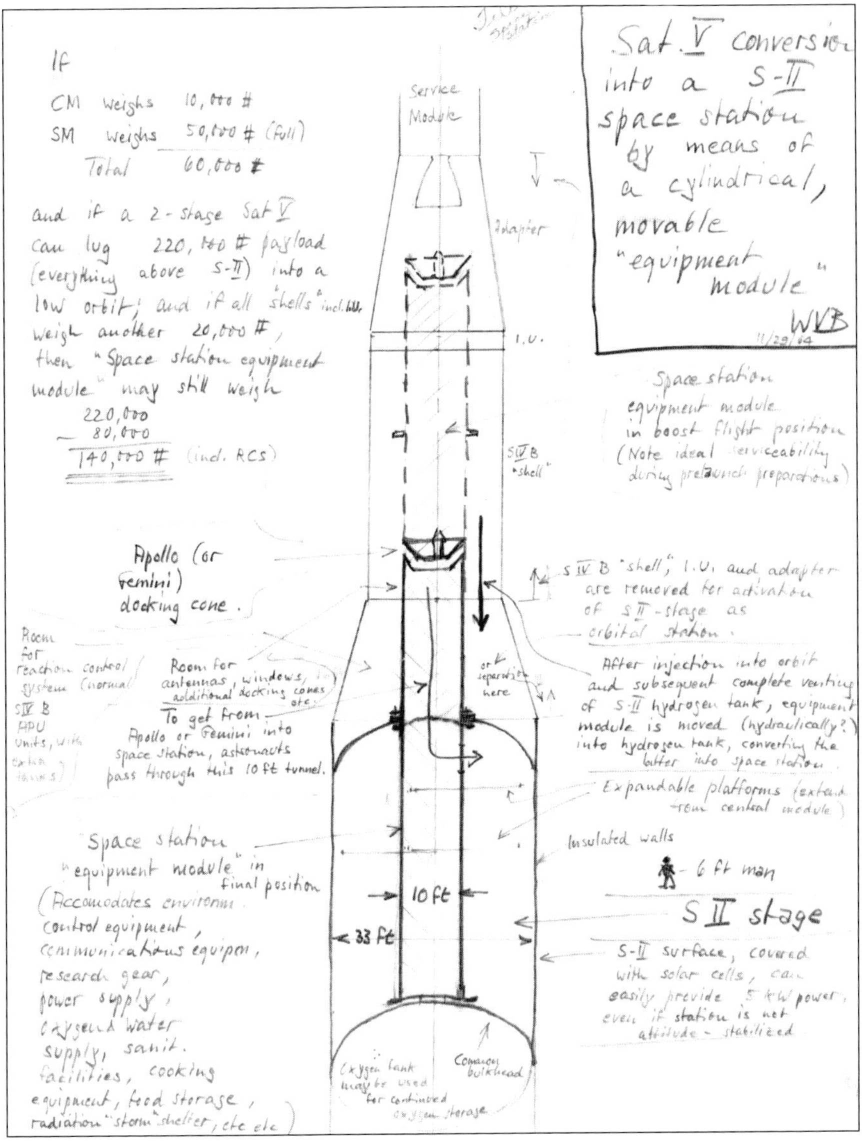

The space station designs of the 1960s tended to borrow hardware from the rockets that would have launched them. It was common for designers to take a stage of a heavy booster, gut it and install inside of it the components of a habitable station. The benefit of this so-called 'wet workshop' approach was obvious. A space station would have to ride into orbit on a multi-stage rocket, anyway. Depicted here, Wernher von Braun's 1964 sketch of a space station based on a conversion of a 'wet' Saturn V second stage. (NASA art)

The Manned Orbital Laboratory's main module wasn't literally a leftover rocket stage, but the US Air Force and McDonnell Douglas scaled it to fit atop a Titan IIIC heavy booster. When NASA developed the Skylab space station starting in the mid-1960s, the agency on Wernher von Braun's advice considered different configurations, each borrowing a different stage of a Saturn rocket. NASA ultimately settled on a leftover Saturn IV aeroshell as Skylab's main module. The station launched into low orbit atop a Saturn V rocket – the last in NASA's inventory – on May 13, 1973. In this photo, workers at McDonnel Douglas's ill-fated Huntington Beach facility handle the hull of a MOL mock-up. Many of those workers would lose their jobs when the Air Force canceled MOL in 1969. (US National Reconnaissance Office photo)

In the early 1960s, the US Air Force considered borrowing NASA's two-layer Gemini space suit for the military astronauts it was recruiting for the Manned Orbital Laboratory program. Instead, the Air Force commissioned the Hamilton Standard Corporation to develop a single-layer suit with extra flexibility. The new suit was supposed to help the MOL's crew squeeze through the hatch between the lab and the Gemini capsule that carried them into orbit. The fully-pressurized suit opened in the front. To enter it, the astronaut had to open and lift the neck ring, release the pressure seal around the neck, pull down the zipper then climb into the suit through the neck-opening. For extra-vehicular work, Hamilton Standard also developed a bulky white thermal suit that an astronaut wore over the blue pressure suit. After the MOL program collapsed, the Air Force donated the special suits to NASA for further evaluation. (US National Reconnaissance Office photo)

The US Air Force hired contractors to build a succession of mock-ups of the Manned Orbital Laboratory for testing and training. The medium-fidelity mock-ups, one of which is depicted here, featured limited functionality but, perhaps most importantly, was the same dimensions as the operational lab and thus allowed astronauts to get a feel for life inside America's planned military space station. MOL actually was fairly roomy by a space explorer's standards. Each of the two crew had around 400 cubic feet of space. Beds were at one end of the lab, storage was at the other. The kitchen and toilet occupied one wall, the other held the workstations, there was no shower. (US National Reconnaissance Office photo)

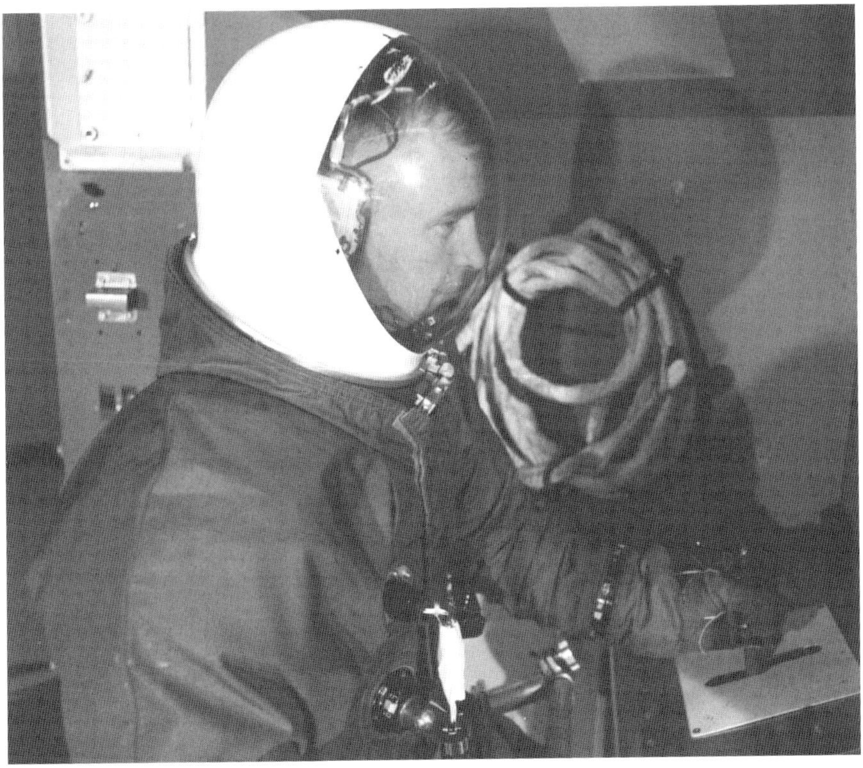

After a year of study, the US Air Force, the US National Reconnaissance Office and the office of US Defense Secretary Robert McNamara concluded that, as a recon platform, the Air Force's Manned Orbital Laboratory held certain advantages over an unmanned spy satellite. Harold Brown, McNamara's head of research and development, emphasized a crew's 'ability to recognize patterns and interpret them in real-time and report the results' as well as 'the ability to point a sensor and provide image-motion compensation so that a very narrow field of view can encompass the area to be examined'. In this photo, a tester wearing one of the unique MOL spacesuits works in a mock-up of the manned station. (US National Reconnaissance Office photo)

To understand potential living conditions in the weightlessness of the Manned Orbital Laboratory, McDonnell Douglas engineers at the company's facility in Huntington Beach, California, submerged the mockup in water and sent in divers. The same method allowed NASA to test spacewalk techniques. The station, like all long-endurance spacecraft, undoubtedly would require occasional external repairs that meant spacewalks by station crew. It was critical to install hatches and handholds in the most efficient positions to minimize the stress on the spacewalker, a lesson NASA learned the hard way during the Gemini missions in the early 1960s. (US National Reconnaissance Office photo)

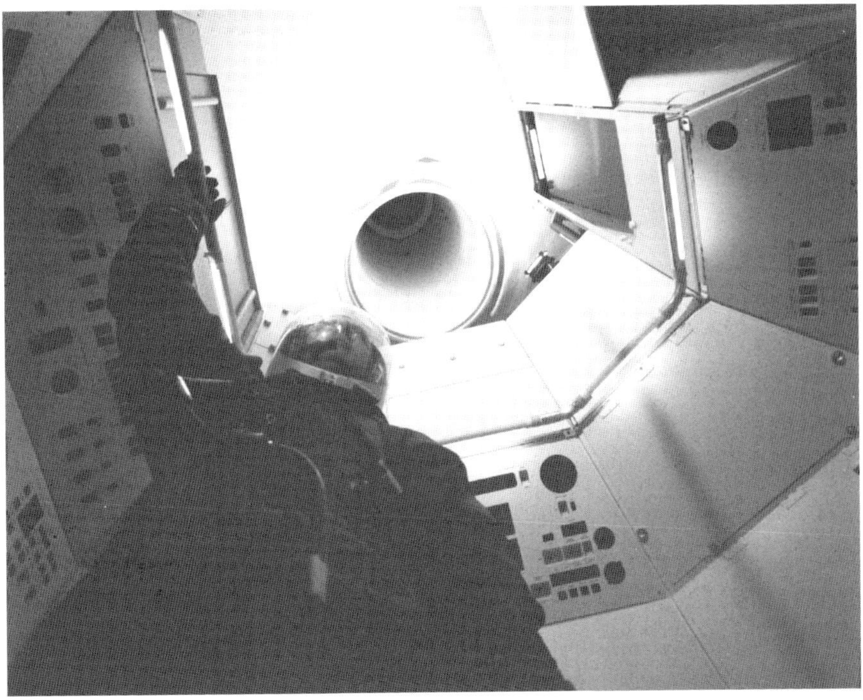

The Manned Orbital Laboratory never made it into orbit, but many of the people who worked on it in the Air Force and McDonnel Douglas went on to work on other space programs – and presumably brought their expertise with them. Likewise, key technologies from the MOL program migrated to other projects. Consider MOL's unique spacesuit. The Air Force anticipated it would be difficult for crew members wearing Gemini-style spacesuits to squeeze through MOL's hatch so station crew would wear the extra-flexible suit that Hamilton Standard Corporation developed. NASA ultimately borrowed elements of the suit's design, although not its color. (US National Reconnaissance Office photo)

PART FOUR

BEYOND JUPITER

A very good rocket,
a very bad rocket and
a prayer for two moonwalkers

Chapter 23

Zvezda

It should come as no surprise that, while the Pentagon was hard at work on the Dyna-Soar, the Kremlin was developing its own, similar space-plane. The Soviets after all had access to the same literature – and the same pool of ex-Nazi talent – that informed Dyna-Soar's design.

The Tupolev design bureau, which was most famous for its work on long-range bombers – air-breathing bombers, that is – starting in the mid-1950s tinkered with a design for a so-called 'rocket plane'. The Tu-136 Zvezda was a manned, hypersonic vehicle with a solid-fuel rocket booster and a 20-ton payload. If the basic vehicle proved viable, Tupolev planned to develop variants for a wide range of missions. There would be a reconnaissance variant, a bomber, an interceptor.

Like Dyna-Soar, Zvezda would be capable of atmospheric flight but, with a bit more thrust, might also achieve a low Earth orbit and even change orbits by way of maneuvering rockets. In the same way that Dyna-Soar had the potential to become the United States' first spacecraft – the means by which the first American would travel to space – Zvezda also could have been the Soviet Union's ticket into space. Dyna-Soar got canceled, of course and when an American punched through the Karman Line for the first time, it was aboard a capsule riding atop a vertical-launching rocket. Zvezda likewise never got off the page and the first Soviet to reach space did so aboard a capsule.

Zvezda's booster was part of the problem. A space-plane carrying its own payload was a heavier lift than a capsule but that wasn't a fatal liability for Dyna-Soar. The US Air Force could count on the heavy rockets von Braun was developing. Saturn, in particular.

The Soviet Union had Korolev, of course. At the same time von Braun was working on Saturn, former political prisoner Korolev was working on his own heavy rocket, N-1.

Saturn's main role was to throw a manned capsule toward the Moon. Boosting Dyna-Soar was a secondary mission and a short-lived one, at that. N-1 likewise was a Moon rocket that also could have boosted a Soviet space-plane.

It's no accident that the United States and the Soviet Union in their respective space programs aimed for similar goals at the same time. Yes, the two superpowers were deliberately racing each other, each spying on the other and launching new programs in pursuit of the same technology they perceived the other to be developing. Moreover, both countries for a time tapped the same wellspring of ex-Nazi expertise. The USA and USSR both were working on a hypersonic space-plane in combination with a superheavy rocket because both countries had read Sänger and Bredt and recruited Peenemünde rocket enthusiasts.

The Americans and Soviets were working with the same conceptual and intellectual raw material, but the Soviets initially had better luck.

The USA and USSR both developed small satellites to fit in the nose cones of rockets that owed their basic technology to von Braun and the V-2. But the Soviets were in a bigger hurry than the Americans and there was less technological perfectionism on the Soviet side so Sputnik 1 got into space before Explorer 1. Both countries likewise aimed to put a man in orbit inside a capsule riding on top of a multi-stage rocket. Again, the Soviets were slightly luckier and slightly faster, in part because the Soviets were less concerned with an elegant re-entry. Yuri Gagarin arrived in space before Alan Shepard did – and nearly burned up on his way home.

But the Americans beat the Soviets to the Moon. Indeed, the Soviet Union never went to the Moon at all, at least, not with a human crew. That seminal feat by the USA – landing people on the lunar surface after a harrowing, four-day journey – all but erased in the popular imagination the USSR's earlier space victories. The Americans owed that feat to Saturn's success … and N-1's failure. Saturn's development wasn't entirely painless but the American rocket never crashed or exploded.

There were hiccups and distractions during Saturn's development. NASA early in the Apollo program toyed with the idea of making at least some portions of the Moon-mission hardware reusable. In the end, almost all of it was single-use. But in an early incarnation of an idea that five decades later would become a cornerstone of efforts to drive down the cost of space launches, Boeing in 1963 drew up concepts for a multiple-use first stage for Saturn rockets.

Instead of launching vertically from a pad, a three-stage Saturn rocket would plug horizontally onto a broad, six-engine airplane with a 105-foot wingspan and an odd, asymmetrically-fitted cockpit. The carrier would take off from a runway like any other plane, then pull up into a vertical climb. The rocket's first stage would fire, dragging the carrier and its poor pilot with it.

The first stage with its attached carrier presumably would have separated from the rocket's upper stages at the usual 220,000 feet, at which point the carrier would spiral down for a horizontal landing, the spent first stage still attached to its centerline – 220,000 feet, or 42 miles, is high. The Karman Line is only slightly higher at 62 miles. In other words Boeing's Saturn-carrier very nearly would have been a spaceplane in its own right. NASA wisely passed on the Boeing proposal.

Von Braun assembled an army of engineers, a mix of NASA employees and agency and private contractors, to design, test and build Saturn. There were two key milestones ahead of the first test launch, each of which tested Von Braun and underscored his strengths as a manager. The first came in 1959 at the Hay-Adams hotel in Washington, DC Von Braun sat across a table with Abraham Silverstein, a manager from NASA headquarters. Von Braun, the ex-Nazi, shared a bottle of Jack Daniel's with the Jewish Silverstein.

According to *The Los Angeles Times*, the two rocketeers eyed each other like spiders. But if there was ideological tension in the room, it wasn't evident. The men stuck to the agenda and the agenda was rocket fuel. Von Braun was determined to stick with tried-and-true kerosene fuel for the second and third stages of a three-stage Saturn.

Silverstein had a better idea. His scientists had been researching liquid hydrogen. It'd never been tried before, but it promised to be much more potent per pound than kerosene. Silverstein made a convincing case, Von Braun relented and eight years later, when the first three-stage Saturn blasted off, the German wrote Silverstein a note of thanks.

The second milestone came in 1964, when George Mueller, the head of NASA's Office of Manned Space Flight, convinced the conservative Von Braun that, rather than separately testing each Saturn component, he should build a whole rocket and test the whole thing all at once. 'It sounded reckless, but George Mueller's reasoning was impeccable,' Von Braun later admitted. So-called 'all-up testing' speeded up Saturn's development by years.

For all the innovation in fuel and testing, when Saturn finally came together as a functional launch vehicle, it featured an entirely conventional configuration – a vertical-launching stack of disposable rocket stages of decreasing diameter, with the payload on top.

After months of ground-testing, the first Saturn V – the model of the superheavy rocket that would take the Apollo 11 astronauts to the Moon – blasted off from Launch Complex 39A at the John F. Kennedy Space Center on November 9, 1967. Generating a staggering 7.5 million pounds of thrust, as much as 125 V-2s, the 363-foot, three-stage rocket carried an unmanned Apollo command and service module 9,800 miles high.

'The whole world could see the awesome sight of the first launch of what is now the largest rocket ever flown,' President Johnson crowed.

It was a feat the Soviets couldn't match. Their N-1 was busy blowing up.

Above: Wernher von Braun discusses the Saturn rocket with President John F. Kennedy at Cape Canaveral, Florida, on November 16, 1963. By then, von Braun's team had completed four Saturn test launches. In stark contrast to the Soviet Union's experience with the similar N-1 rocket, none of the Saturns failed. Indeed, no Saturn ever failed during the type's 32 launches between 1961 and 1975. (US National Archives photo)

Opposite above: NASA succeeded in luring Wernher von Braun away from the US Army by promising to fund his ambitious Saturn super-booster. Von Braun and his engineers at the Marshall Space Flight Center had already drawn up designs for an enormous rocket first stage with F-1 engines burning a mix of cryogenic liquid oxygen and kerosene. Exactly how many engines depended on the model. The ultimate Saturn V packed five F-1s. In 1964, von Braun and his team – now part of NASA – began testing the first-stage fuel tank, pictured here. It was critical for von Braun to understand the behavior of the fuel during launch. Too much sloshing could throw off the rocket's trajectory and doom a mission. (NASA photo)

Opposite below: In July 1964, Douglas Aircraft delivered the first test hardware for the Saturn V's third stage from Long Beach, California, to NASA's Marshall Space Flight Center in Alabama. The S-IVB third stage measured 21.5 feet in diameter and 58 feet in length. Its J-2 engine was capable of producing a whopping 225,000 pounds of thrust. Pictured here in Long Beach, is the fully assembled S-IVB that was destined to support the Apollo 8 mission. (NASA photo)

The Apollo 4 mission on November 9, 1967 was the first test flight of Wernher von Braun's Saturn V superheavy booster and the first rocket test at the new Launch Complex 39A at the John F. Kennedy Space Center in Florida. NASA originally scheduled Apollo 4 for late 1966, but delayed the mission to November 1967 owing to problems with North American Aviation, the manufacturer of the Saturn. Apollo 4 was also the first Apollo flight after a safety stand-down that NASA put in place after the Apollo 1 fire on January 27, 1967. The accidental blaze killed Apollo 1 astronauts Gus Grissom, Ed White and Roger Chaffee. (NASA photo)

Apollo 8, the first crewed Saturn V mission, was scheduled to launch from NASA's John F. Kennedy Space Center on December 21, 1968. First, workers had to assemble the rocket. Pictured here, crews erect the S-IC first stage in Kennedy's vehicle assembly building. (NASA photo)

Apollo 8 was the first manned mission for Wernher von Braun's Saturn V, which remained the world's most powerful space launch vehicle through at least 2021. The Saturn V that boosted Apollo 8 generated more than seven million pounds of thrust shortly after lift-off from Launch Complex 39A at NASA's John F. Kennedy Space Center in Florida. (NASA photo)

Chapter 24

NASA

The first N-1 to break cover was actually a full-scale, non-flying mock-up that an NRO satellite spotted on a launch pad at the new Baikonur Cosmodrome in Kazakhstan on November 25, 1967. Where the Americans assembled their heavy rockets in tall hangars then hauled them out to their launch pads on special crawlers, the Soviets assembled and transported their rockets horizontally and raised them to vertical at the launch pad. That took practice and the bigger the rocket, the more practice crews needed, hence the N-1 mockup. It was never meant to fly. Crews lowered the rocket after a couple of weeks and the first N-1 launch wouldn't take place until February 1969.

But the NRO – and, by extension, the CIA and NASA – didn't know that. The N-1 erection test in late 1967 may have compelled NASA to accelerate some aspects of the Apollo program. Specifically, NASA opted to go ahead with the Apollo 8 Moon mission, even though the Apollo program's lunar lander wasn't yet ready. Instead of landing on the Moon, the Apollo 8 astronauts Frank Borman, Jim Lovell and William Anders rode their Saturn *around* the Moon and then back to Earth.

In fact, the N-1 wasn't about to threaten America's head start getting to the Moon. Four catastrophic failures of N-1s in tests between 1969 and 1972 doomed the Soviet Union's effort to follow the Americans to the lunar surface. The explosions as a side effect also grounded the Tu-136 concept. Soviet leadership had approved the N-1 project way back in 1960 and, four years later, designated the superheavy rocket as the booster for a future manned Moon landing. But that prioritization did not result in the N-1 program getting the people and resources it needed to meet the government's arbitrary deadline for a first test launch in the third quarter 1967. While American resources concentrated inside the Apollo program and its own Saturn booster, Soviet resources remained hopelessly spread out across several competing design bureaus, each with its favorite projects.

In retrospect, the years-long bureaucratic battle between the US armed services and the agency that became NASA – the fight that determined who would lead the American space program – was decisive in the United States beating the Soviet Union to the Moon. With one agency in charge, the Americans were able to focus resources and because NASA didn't solely depend on the patronage of one or a handful of powerful and temperamental men, its expanding power was enduring. That translated into clearer goals, less waste, rockets that usually worked. and ambitious space missions that stood a decent chance of succeeding.

It's not that von Braun didn't face internal friction as he designed and built the world's most powerful rocket. Both NASA and the Air Force briefly tinkered with alternatives to Saturn. But those boosters – in particular, NASA's Nova – were insurance policies against Saturn's failure more than they were *rivals* of von Braun's own giant rocket. Once it was clear Saturn worked, the other heavy booster projects faded away.

That NASA's mission was scientific rather than military if anything bolstered its independence and thus its effectiveness. The United States' fractious politics in the 1950s and 1960s – the era of civil rights, political assassination, war protest and the rise of future impeached President Richard Nixon – nudged NASA this way and that. But *only* nudged. The space agency proved to be stubbornly nonpartisan. President Dwight Eisenhower, a Republican, created the agency, albeit begrudgingly. President John F. Kennedy, a Democrat, directed the agency to aim for the Moon, his vice president and successor President Lyndon Johnson doubled down.

Nixon was a Republican, and a divisive one. But under his watch, NASA finally achieved the goal Kennedy had set for it and which Johnson had sustained. It was a goal that was a natural extension of the mission Eisenhower had ascribed to NASA when, a year after Sputnik, he reluctantly signed the law that created the agency. When the Americans beat the Soviets to the Moon in July 1969, it helped to lock into place the expectation that space was for science and for peace. It's perhaps fortunate that, by then, von Braun worked for NASA and not for the US Army.

By the time of the Moon landing, von Braun had abandoned his most militaristic concepts. After all, his sponsor wasn't really interested. NASA had succeeded where the Soviets had failed, in large part because the Soviets didn't have their own NASA. The Soviets never resolved their own bureaucratic conflict. Infighting wasted time. Dead-end programs burned money. Institutional instability undermined long-term projects. Korolev's death in 1966 underscored this dysfunction. The N-1 rocket, the key to the Soviet Moon push, more or less died with him.

Korolev had been diagnosed with colon cancer in 1965, but it wasn't the cancer that killed him. His health never truly recovered from his time in the gulag. Weary and fragile, he died on the operating table at a Moscow hospital on January 14. He was just 59 years old. His death only exacerbated the crisis inside OKB-1 and the N-1 program. In the spring of 1967, with the deadline for the first N-1 launch looming and progress on the giant rocket lagging, Soviet authorities ordered cadets at the national military engineering school to cut short their studies, graduate early and report to Baikonur. It didn't help. The deadline for the first N-1 launch came and went. Two years later, the rocket finally was 'ready'.

In fact, the Central Research Institute for Machine Building in Moscow – an internal think tank for the Soviet space industry – assessed the N-1 and concluded the first test launch had just a one-percent chance of success. The institute wasn't wrong. Around noon on February 21, 1969, the first N-1 blasted off, its 30 first-stage engines roaring and spewing smoke as they burned kerosene and liquid oxygen. A minute later, the engines began failing. The N-1 bled speed, stalled out nine miles over the Earth's surface ... and fell. Three minutes after launch, the giant rocket smashed into the ground 32 miles from the pad.

Investigators later blamed the failure on a combination of a faulty telemetry system that relayed nonsensical commands from the ground to the rocket and loose pipes that leaked fuel into the rocket. The former prematurely shut off some of the engines, the latter resulted in an out-of-control fire.

Vasili Mishin, Korolev's successor as the *de facto* head of the Soviet space program, was happy that the N-1 at least got off the pad. He pushed for a second test on July 3. Boris Chertok, who was responsible for the N-1's control system, warned a member of the state commission overseeing the Soviet space program there wasn't enough time to fix the problems with the telemetry. Chertok was warned not to tell the full commission that he couldn't fix the telemetry system. Panic was setting in as the Americans pulled further ahead in the race to the Moon. Realistic appraisals of the faltering Soviet Moon program were fast becoming politically unacceptable in official circles. Better to lie and take your chances the truth wouldn't make itself known.

The second N-1 test went ahead with some minor fixes to the internal fittings. As a precaution, thousands of non-essential personnel evacuated Baikonur. Other essential personnel dug trenches they hoped would protect them from an explosion.

The second N-1 failed even faster than the first. It was just 300 feet over the pad when segments of its first stage separated and plummeted to the ground. The N-1 lost thrust, hovered for a moment then fell. Personnel manning a fuel station dived into their trench as the fireball rolled over them. 'I saw without exaggeration the end of the world,

The Soviet space program rolled out a mockup of its N-1 superheavy rocket at the new Baikonur Cosmodrome in Kazakhstan on November 25, 1967. Where the Americans assembled their heavy rockets in tall hangars then hauled them out to their launch pads on special crawlers, the Soviets assembled and transported their rockets horizontally and raised them to vertical at the launch pad. (RKK Energia photo)

and not in a nightmare, but while fully awake and standing right next to it,' Lieutenant Colonel Semen Komarovsky said.

Investigators couldn't settle on a single clear cause, although one theory – that debris entered a fuel pump – was politically acceptable. The NRO had a satellite in position to watch the launch.

The Soviet Union's Moon program was in a state of collapse, and the Americans knew it. Apollo 11 landed Neil Armstrong and Buzz Aldrin on the Moon 17 days later on July 20, 1969.

There was still value in a second-place finish. The Soviets scheduled a third N-1 test for June 27, 1971 but it also ended in an

Soviet leadership had approved the N-1 project way back in 1960 and, four years later, designated the superheavy rocket as the booster for a future manned Moon landing. But that prioritization did not result in the N-1 program getting the people and resources it needed to meet the government's arbitrary deadline for a first test launch in the third quarter of 1967. An N-1's third stage with its four engines is depicted here. (RKK Energia photo)

explosion and so did a fourth test on November 23, 1972. The Central Committee of the Communist Party of the Soviet Union fired Mishin in May 1974.

His replacement Valentin Glushko, an old rival of Korolev, was eager to erase his competitors' legacies. He ended existing lunar efforts, ordered engineers to dismantle any leftover N-1s and got to work designing a new Moon program from scratch – one that included a fantastical lunar colony. But in 1974 even America's Moon program was ending after six successful landings. The space race had new goalposts. Glushko's colony concept went nowhere.

Embarrassed, the Soviet Union buried the details of the N-1 program. Records finally began surfacing thanks to the USSR's belated *glasnost*, or 'openness', policy in the late 1980s. As recently as the 2010s, previously unreleased photos and videos of N-1 development and testing still were coming to light.

The Tu-136 Zvezda was a manned, hypersonic vehicle with a solid-fuel rocket booster and a 20-ton payload. If the basic vehicle proved viable, Tupolev planned to develop variants for a wide range of missions. There would be a reconnaissance variant, a bomber, an interceptor. The Soviet Union ultimately never built the Tu-136, but it did build several prototypes of the similar Tu-139, an unmanned, air-breathing reconnaissance drone. The 92-foot-long Tu-139s, one of which is pictured here, flew test sorties in the late 1960s and early 1970s, but never evolved into an operational vehicle. (Tupolev photo)

Like the Dyna-Soar, Zvezda would be capable of atmospheric flight but, with a bit more thrust, might also achieve a low Earth orbit and even *change* orbits by way of maneuvering rockets. In the same way that Dyna-Soar had the potential to become the United States' first spacecraft – the means by which the first American would travel to space – Zvezda also could have been the Soviet Union's ticket into space. Pictured here, one of the prototypes of the Tu-139, an air-breathing reconnaissance drone that shared design elements with the Tu-136. (Tupolev photo)

The N-1 boasted a whopping 30 engines on its first stage. Problems with the complex propulsion system, among other faults, resulted in all four N-1 test launches failing between 1969 and 1972. The crashes effectively ended the Soviet effort to land people on the Moon. (RKK Energia photo)

The first N-1 to break cover was actually a full-scale, non-flying mockup that a US National Reconnaissance Office satellite spotted on a launch pad at the new Baikonur Cosmodrome in Kazakhstan on November 25, 1967. Pictured here, that mock-up and the first functional N-1, which exploded during its uncrewed test launch on February 21, 1969. Investigators later blamed the failure on a combination of a faulty telemetry system that relayed nonsensical commands from the ground to the rocket and loose pipes that leaked fuel into the rocket. The former prematurely shut off some of the engines, the latter resulted in an out-of-control fire. Vasili Mishin, Korolev's successor as the *de facto* head of the Soviet space program, was happy that the N-1 at least got off the pad. (RKK Energia photo)

The US National Reconnaissance Office had a satellite in position to watch the second N-1 launch on July 3, 1969. The launch failed even faster than the first did. The N-1 was just 300 feet over the pad when segments of its first stage separated and plummeted to the ground. The N-1 lost thrust, hovered for a moment then fell. Personnel manning a fuel station dived into their trench as the fireball rolled over them. 'I saw without exaggeration the end of the world, and not in a nightmare, but while fully awake and standing right next to it,' Lieutenant Colonel Semen Komarovsky said. (US National Reconnaissance Office photo)

The United States' clear victory over the Soviet Union in what in retrospect obviously was the final lap of the space race actually helped to reduce tensions between the two orbital powers. There was, in the 1970s, a period of detente as NASA and the Soviet space program cooperated more than they competed. In 1972, a group of Soviet educators visited NASA's John F. Kennedy Space Center in Florida, as seen in this photo and in July 1975, an Apollo capsule, boosted by the last of Wernher von Braun's Saturn rockets, docked with a Soviet Soyuz capsule. (US National Archives photo)

Chapter 25

Three men to the Moon

On the morning of July 16, 1969, astronauts Neil Armstrong, Buzz Aldrin and Michael Collins sat 363 feet in the air, squeezed shoulder-to-shoulder in the command-module capsule at the tip of a Saturn V rocket resting on the pad at Launch Complex 39A at John F. Kennedy Space Center.

The three stages beneath them together were capable of generating, at their peak, nine million pounds of thrust. Enough to boost the rocket's third stage – containing the command and service module and the lunar lander – roughly 230,000 miles to the Moon.

It was hot and humid but the weather was clear. Conditions were looking good for the 9:32 am launch. Half a million people packed the roads and beaches surrounding the cape. 'People brought their children, their folding chairs and their binoculars,' photographer David Burnett recalled. Many had camped overnight in and around Titusville, the town right across the Indian River from Launch Complex 39A. 'History observed in shirt-sleeves and bikinis,' CBS News anchor Walter Cronkite mused in his broadcast.

No fewer than 4,000 reporters crowded special stands and platforms: cameras rolling, tape recorders spooling, notepads in hand. Other stands supported hundreds of US lawmakers and foreign dignitaries and the US vice president, Spiro Agnew.

President Nixon watched from the White House. His speechwriter William Safire had penned a short announcement for Nixon to read to the public in the event the Apollo hardware failed, stranding Armstrong and Aldrin on the Moon. 'Every human being who looks up at the Moon in the nights to come will know that there is some corner of another world that is forever mankind,' the speech read, in part.

Hundreds of millions of people all over the world sat in front of television sets, hoping to glimpse the explosive first step in a journey that should, God and von Braun willing, safely land the first human explorers on the Moon. Von Braun sat in the complex's launch control center having flown in by helicopter. As a licensed pilot, he'd even taken the controls for part of the flight.

The reverse countdown, a holdover from Fritz Lang's science-fiction film, *Woman in the Moon*, on which von Braun's long-ago mentor Hermann Oberth had consulted, steadily subtracted. Von Braun closed his eyes, bowed his head and silently recited the Lord's Prayer. *Our Father, who art in Heaven, hallowed be Thy name; Thy kingdom come; Thy will be done on Earth as it is in Heaven.*

The five F-1 engines in the Saturn V's mammoth first stage ignited with a thunderclap roar. Smoke and flame billowed across the pad, birds scattered and thousands of people gasped.

Nixon's consolatory speech was unnecessary. On the morning of July 20, Armstrong and Aldrin, riding in the crablike lunar lander *Eagle*, safely touched down on the Moon. Nixon, watching from the White House, reacted with a simple, 'Hooray!'

Von Braun had traveled to mission control in Houston to watch. It was, he recalled, 'a pretty emotional moment'. If the ex-Nazi's thoughts were with the many thousands of people who died to make the moment possible, he never mentioned it.

In August 1970, NASA named von Braun as the agency's new deputy associate administrator for planning and transferred him to Washington, DC. The people of Huntsville organized a parade to say goodbye to their longtime local hero. Von Braun retired from NASA two years later in July 1972. He accepted an offer to become vice president for engineering and development at Fairchild Industries, a San Antonio aerospace firm. The move came with a big pay raise.

At Fairchild, von Braun reportedly earned around a million dollars a year. There were perks, too, including a car and driver.

NASA was less than thrilled. 'Dr. von Braun's decision to retire from NASA is a source of great regret to all of us at the agency,' NASA administrator James Fletcher said. 'For more than a quarter of a century, he has served the United States as the leader in space-rocket development.'

There was no convincing von Braun to stay. The Apollo 11 mission represented a high-water mark for US space funding. The USA had beat the USSR to the Moon. With Korolev dead and buried, the Soviets visibly were giving up on their most ambitious space projects. It perhaps was inevitable that NASA's budget would begin to contract, too. Nonetheless, the spending cuts were demoralizing to von Braun. He wanted to build on the Moon mission and send astronauts onward to Mars but the 1971 budget cycle made it clear that a mission to Mars wasn't going to happen in his lifetime. In 1970 the White House and Congress had lavished \$4 billion on NASA. To get to Mars by 1981, von Braun projected the budget would need to rise to at least \$7 billion. Nine billion dollars would have been better.

But the 1971 budget was only slightly more than \$3 billion, or \$22 billion in 2021 values. NASA axed the last two planned Apollo missions – Apollo 18 and 19 – and delayed a plan to launch Skylab, America's first space station. To von Braun, it felt like the bottom was falling out of the space program. He didn't see the situation turning around any time soon. 'I believe there may be too many people in NASA who at the moment are waiting for a miracle,' he told historian Logsdon. 'They believe we are entitled to this kind of a thing forever, which I gravely doubt.'

Von Braun's team oversaw the completion of 32 Saturns at a cost of \$6.4 billion, which is \$27 billion in 2021 values. Besides launching all 17 Apollo missions – the last in December 1972 – a Saturn V, leftover from the cancellation of the last two Apollo flights, boosted Skylab into orbit in May 1973. Three Saturn IBs flew three astronauts apiece to the station between May and November 1973. And that was it. Von Braun's Saturn superbooster was, and as of mid-2021 still is, the only rocket ever to carry a human being beyond a low Earth orbit. At the end of 1973, there were no more Saturns and no new, super-ambitious manned space missions to justify a production restart.

Apollo 9 from March 3 to 13, 1969, was the Moon program's third manned mission. During their 10 days in low Earth orbit, astronauts James McDivitt, David Scott and Rusty Schweickart tested key aspects of a lunar landing and, for the first time, crewed up the lunar lander and proved the lander could leave the command module, maneuver back to it and dock safely. The mission also represented another trial of the Saturn V booster. Wernher von Braun observed the launch in Houston alongside Vice President Spirow Agnew, representing the incoming administration of President Richard Nixon. (NASA photo)

At 9:32 am on July 16, 1969, a 363-foot Saturn V rocket launched from Launch Complex 39A at NASA's John F. Kennedy Space Center in Cape Canaveral, Florida. Aboard the Apollo 11 spacecraft on the Saturn's third stage were astronauts Neil Armstrong, the mission's commander, plus command-module pilot Michael Collins and lunar-module pilot Buzz Aldrin. Wernher von Braun watched the launch from the nearby launch control center. As the countdown to launch neared zero, von Braun – never a particularly pious or superstitious man – nonetheless bowed his head and silently recited the Lord's Prayer. (NASA photo)

The Saturn V rocket carrying the Apollo 11 crew lifts off from NASA's John F. Kennedy Space Center on July 16, 1969. Mission commander Neil Armstrong and lunar-module pilot Buzz Aldrin four days later would safely land on the Moon's surface while command-module pilot Michael Collins orbited overhead. Armstrong and Aldrin gathered samples of lunar material and set up scientific instruments that transmitted back to Earth data about the lunar environment. More importantly, they demonstrated to the whole world – and the Soviet Union, in particular –that the United States possessed the technological prowess to land people on the Moon before any other country. (NASA photo)

Apollo 11 astronaut Buzz Aldrin stepped onto the surface of the Moon moments after mission commander Neil Armstrong had exited the *Eagle* lander. Technically, that made Aldrin the second man to set foot on the Moon – a label that, for the next five decades, he bitterly rejected. 'Is it really necessary to point out to the crowd that somebody else was first when we all went through the same training, we all landed at the same time and all contributed?' he told *National Geographic* in 2016. Aldrin is pictured here inside the lunar module. (US National Archives photo)

NASA's attention bifurcated. From the mid-1970s on, the agency simultaneously designed and deployed robotic probes for surveying Earth's neighbors – Mars, Venus, Saturn, Jupiter. Meanwhile, NASA poured billions into building a bulky, partially reusable orbiter for low-Earth missions: the ill-fated Space Shuttle – the most failure-prone, dangerous spacecraft in history.

Above: Astronaut Buzz Aldrin descends the lunar module ladder on the Moon on July 20, 1969. Stepping onto the Moon's surface behind mission commander Neil Armstrong, Aldrin looked around and uttered two words that described his own soul as accurately as they did the lunar landscape. 'Magnificent desolation.' 'No oxygen, no life,' Aldrin expounded later, 'just the lunar surface that hasn't changed for thousands of years – and the blackness of the sky.' Aldrin, then an Episcopalian, had brought wine and communion wafers with him into space. In later years, his religious philosophy would evolve into what he described as 'a cosmic sense of a greater power involved in the creation of the universe'. (US National Archives photo)

Opposite above: Apollo 11 astronaut Buzz Aldrin stands beside the Passive Seismic Experiments Package on the surface of the Moon on July 20, 1969. Another instrument called the Laser Ranging Retro Reflector also is visible, along with a US flag, a television camera and the Apollo Lunar Surface Close-up Camera and the lunar module that Aldrin steered down to the Moon's surface. (NASA photo)

Apollo 11 astronauts Neil Armstrong, Buzz Aldrin and Michael Collins traveled to the Moon inside the cramped command module. Once in orbit around the Moon, Armstrong and Aldrin transferred to the lunar module for their trip down to the lunar surface. Every stage of the mission was risky. President Richard Nixon's speechwriter William Safire penned a short announcement for Nixon to read to the public in the event the Apollo hardware failed, stranding Armstrong and Aldrin on the Moon. Pictured here, Armstrong inside the lunar module shortly after his first moonwalk on July 20, 1969. (US National Archives photo)

It was far from a foregone conclusion that NASA's manned mission to the Moon would succeed. Apollo 11 easily could have failed at multiple junctures, starting with the launch. Wernher von Braun's Saturn V super-booster had proved highly reliable by July 1969 – five had launched without incident. Contrast that with the Soviet experience – the USSR's own superheavy booster, the N-1, *never* safely launched. When the Apollo 11 astronauts landed on the surface of the Moon, von Braun recalled experiencing 'a pretty emotional moment'. Upon their safe return to Earth eight days after launch, Neil Armstrong, Buzz Aldrin and Michael Collins packed into a mobile quarantine unit aboard the aircraft carrier USS *Hornet*. They spent nearly three days in the quarantine trailer on *Hornet* before transferring to a bigger isolation unit on land. President Richard Nixon embarked on *Hornet* to greet the astronauts. (US National Archives photo)

The Apollo 11 mission presented NASA officials with a long list of unknowns. While Marshall Space Flight Center director Wernher von Braun worried about the safety and performance of the Apollo 11 rocket, mission medical officer Charles Berry concerned himself with the prospect, however remote, that the Apollo 11 crew might carry Earth diseases into space and thus risk their health while Moonbound or, alternatively, bring back space diseases to Earth … and risk a global plague. Apollo 11 astronauts Neil Armstrong, Buzz Aldrin and Michael Collins spent three weeks in isolation prior to launch and three weeks in quarantine after leaving the Moon. President Richard Nixon almost wrecked Berry's quarantine plan when he insisted on meeting with the astronauts right before liftoff. Berry begged the president to reconsider, and he reluctantly did so. Nixon finally got his chance to pal around with Armstrong, Aldrin and Collins as the three sat in post-mission quarantine in a trailer aboard the aircraft carrier USS *Hornet*. (US National Archives photo)

Above: A helicopter flying from the US Navy aircraft carrier USS *Hornet* zeroed in on the Apollo 11 command module following its safe splashdown in the Pacific Ocean on July 24, 1969. A diver swam to the capsule and tossed in sterile isolation suits and gas masks. Neil Armstrong, Buzz Aldrin and Michael Collins spent a few days in a quarantine trailer aboard *Hornet* then flew to Houston, where they again climbed into their trailer. The three-week post-mission quarantine eventually expanded to include 20 NASA support staff. Armstrong even celebrated his 39th birthday in quarantine, complete with a cake and well-wishers. All the worry over Moon illnesses ultimately proved unfounded. (US National Archives photo)

Opposite above: Wernher von Braun's Saturn rockets worked nearly flawlessly. Under a series of NASA contracts with von Braun closely supervising, Boeing, North American Aviation, Douglas Aircraft and IBM together built 32 Saturns. All 32 missions using Saturn boosters succeeded. Arguably NASA's biggest failure of the era, the Apollo 13 Moon mission, had nothing to do with Saturn. The Apollo 13 command module suffered electrical problems mid-flight in April 1970. The crew aborted the lunar landing and focused on safely returning to Earth. Moon missions with Saturn boosters resumed nine months later with Apollo 14. Pictured here, the Apollo 14 crew of Alan Shepard, Stuart Roosa and Edgar Mitchell document their activities while on the Moon between February 5 and 6, 1971. (US National Archives photo)

Wernher von Braun relaxes following the successful launch of the Apollo 11 Moon mission on July 16, 1969. Apollo 11 represented a high-water mark for US space funding. It perhaps was inevitable that NASA's budget would begin to contract but the spending cuts still were demoralizing to von Braun. He wanted to build on the Moon mission and send astronauts onward to Mars but the 1971 federal budget cycle made it clear to von Braun that a mission to Mars wasn't going to happen in his lifetime. (NASA photo)

By 1970, Americans were frequent visitors to the Moon. The space race was won. In August 1970, NASA 'promoted' Wernher von Braun from his current position leading the agency's Marshall Space Flight Center in Huntsville to deputy associate administrator for planning at NASA headquarters in Washington, D.C. Von Braun went willingly. Von Braun's biographer Michael Neufeld cited two major reasons: one, his wife Maria was unhappy in Huntsville, two, the hard-working von Braun was tired. The new job promised to be less demanding than his previous role designing the world's biggest firecrackers. Shortly before von Braun's departure, Alabama governor Albert Brewer unveiled a plaque commemorating the German's long career in Alabama. From left to right at the ceremony, von Braun's daughter Iris, his wife Maria, US senator John Sparkman, Brewer, von Braun's son Peter and his daughter Margrit. (US National Archives photo)

Chapter 26

Von Braun's death

In mid-1973, von Braun renewed his pilot's license. The process required a physical and the X-rays found something around his left kidney. Von Braun procrastinated. It wasn't until a month later that he saw a specialist when he was told, in no uncertain terms, that he needed to lose the kidney, immediately. The surgery was successful and, for a while, von Braun resumed his normal life, traveling on behalf of Fairchild to promote the company's A-10 close-air-support plane and other programs.

A little over a year later, von Braun discovered he was bleeding from his rectum. In August 1975, he went to Johns Hopkins hospital in Baltimore to have some of his colon removed. There was a second surgery to clean up abscesses from the first. Von Braun spent a month recovering at Johns Hopkins.

In March 1976, von Braun celebrated his 64th birthday. By then he was walking with a cane. Perhaps anticipating his impending death, he sat down to write an 82-page treatise on science and religion that he submitted to a Lutheran Church conference. It was far from profound. After some unoriginal ruminations on the need for faith to provide context for science, the paper 'becomes a compilation of recent space advocacy speeches,' according to von Braun's biographer Michael Neufeld.

In mid-May 1976, von Braun suddenly collapsed. His family checked him into Alexandria Hospital in Virginia. As he lay dying, his friends pulled strings in an effort to get von Braun a presidential commendation.

The administration of President Gerald Ford demurred. 'Sorry, but I can't support giving the medal of freedom to a former Nazi whose V-2 was fired into over 3,000 British and Belgian cities,' White House advisor David Gergen wrote in reply. However Jimmy Carter, the former Georgia governor who beat Ford in the 1976 presidential election, was more fond of von Braun and he approved the German for a medal.

The award from Carter wasn't von Braun's last honorific. That came in the form of correspondence from Albert Speer, Hitler's former armaments chief. Speer, who had spent two decades in prison for war crimes, sent von Braun a copy of his memoir. Von Braun thanked Speer in a note. 'How often I thought of you during those 20 long years when so much was happening in my life,' von Braun wrote.

He died on June 16, 1977.

The Washington Post's obituary was generous. 'His extraordinary service to Nazi Germany was a source of considerable inner turmoil after World War II,' J.Y. Smith wrote. 'Thus, he was constantly seeking uses for space technology that he hoped would make the world a less dangerous place.'

It was no secret that von Braun had, at the very least, been cozy with Nazis. As the hyper-patriotism of the 1950s ebbed, more and more people – be they film critics, NASA managers or White House aides – grew more comfortable speaking, in the same sentence, von Braun's name and the word 'Nazi'. But the true extent of von Braun's association with National Socialism wasn't widely known until the mid-1980s, when a journalist named Linda Hunt filed under the US Freedom of Information Act a fat stack of requests for Operation Paperclip records.

As Hunt's requests piled up, the US Army first tried to exploit a loophole in order to avoid handing over the files. The service couldn't deny Hunt, but it could charge her for the cost of photocopying the documents she requested. The Army handed her a bill for $240,000. She appealed and the Army ultimately gave up trying to keep her

Skylab's launch atop NASA's last remaining Saturn V rocket on May 14, 1973, depicted here, was in many ways the last of Wernher von Braun's achievements in space, as he for years had lobbied for the United States to establish an enduring presence in orbit. But von Braun retired from NASA in 1972, meaning he observed Skylab's launch the same way millions of other civilians did – in the press. Skylab actually represented a departure from one of von Braun's favorite space-station proposals. As early as 1964, he had proposed what he called a 'wet lab' architecture for a future space station. Under his wet lab concept, the second stage of a Saturn V – depleted of fuel after its launch – would become the outer hull of a space station. The equipment for the station would ride into orbit in the third stage. Once the second-stage hull was in orbit, the equipment in the third stage would slide into the hull via a hatch. For Skylab, NASA ultimately abandoned von Braun's wet lab design in favor of a 'dry lab' – that is, an intact, single-piece station that rode into orbit as the upper stages of a Saturn V. (NASA photo)

and other journalists away from the records. Starting in the spring of 1985, the Paperclip files began to flow into the public domain. Hunt summarized some of the records in the April 1985 issue of *Bulletin of the Atomic Scientists*.

The New York Times got its hands on the same records. 'American intelligence officials concealed the Nazi records of hundreds of former enemy scientists to try to get them into the United States after the Second World War, contrary to a presidential order and against the objections of the State Department,' *Times* reporter Ralph Blumenthal wrote.

The Paperclip files revealed something about von Braun that even his biggest critics didn't know. According to the records Hunt and the *Times* reviewed, some Army official assessing the Third Reich's ranks of rocketeers after the Second World War weighed von Braun's wartime deeds and came to a conclusion – one that compelled some more senior official to intervene on the brilliant young scientist's behalf and clear him for recruitment. Von Braun, the junior official wrote, posed 'a potential security threat'.

Thomas Stafford, the American commander of the Apollo Soyuz Test Project, and docking module pilot Donald Slayton with Soyuz food tubes during the rendezvous between a NASA Apollo capsule and a Soviet Soyuz capsule on July 17, 1975. The Apollo Soyuz Test Project was the final NASA mission in which Wernher von Braun played any direct role. The Apollo capsule in the 1975 rendezvous boosted into low Earth orbit atop NASA's last Saturn V booster, which von Braun's Marshall Space Flight Center developed prior to von Braun leaving NASA in 1972. NASA photo

Sources

'American Reactions to Crisis: Examples of Pre-Sputnik and Post-Sputnik Attitudes and of the Reaction to Other Events Perceived as Threats,' International Affairs Seminars of Washington, October 15-16, 1958

'Announcement of the First Satellite', *Pravda*, October 5, 1957

'A Brief History of Animals in Space', by Tara Gray, NASA, 1998

'Course Recorded: Navy Picks Up Radio Signals – 4 Report Sighting Device', by Walter Sullivan, *The New York Times*, October 5, 1957

'Creating a Memory of the German Rocket Program for the Cold War', by Michael Neufeld, in *Remembering the Space Age*, edited by Steven Dick, NASA History Division, 2008

Crossbow and Overcast, by James McGovern, William Morrow, 1964

'The Disney – von Braun Collaboration and Its Influence on Space Exploration', by Mike Wright, in *Selected Papers from the 1993 Southern Humanities Conference*, Southern Humanities Press, 1993

'Dr. Wernher von Braun, 65, Dies', by J.Y. Smith, *The Washington Post*, June 18, 1977

'Dyna-Soar plus Titan III', by J.S. Butz, Jr., *Air Force*, February 1, 1962

'Former G.I.s Spill Secrets Of WWII PoW Camp,' by Pam Fessler, National Public Radio, August 18, 2008

Fort Hunt Oral History: P.O. Box 1142 interview with Werner Moritz, by Brandon Bies, US National Park Service, July 23, 24, 2007

'Fort Hunt's Quiet Men Break Silence on WWII', by Petula Dvorak, *The Washington Post*, October 6, 2007

'The German Space Mirror', *Life*, July 23, 1945

Glennan's NASA Diary, Smithsonian National Air & Space Museum

A Grand Guy: The Art and Life of Terry Southern, by Lee Hill, Harper Collins, 2001

'History and the Current Status of the Russian Early-Warning System', by Pavel Podvig, *Science and Global Security*, January 2002

A History of the Dora Camp, by Andre Sellier, Rowman & Littlefield, 2003

'Industries in Space to Benefit Mankind: A View Over the Next 30 Years', Rockwell International and NASA, 1977

Interview with Herbert York, American Institute of Physics, September 24, 1980

'*Mars Project*: Wernher von Braun as a Science-Fiction Writer', by Michael Neufeld, Smithsonian National Air & Space Museum, January 22, 2021

James Van Allen: The First Eight Billion Miles, by Foerstner, Abigail, University of Iowa Press, 2007

'Largo veteran's photos recall the horror of the Gardelegen Massacre', by Piper Castillo, *Tampa Bay Times*, November 8, 2013

Making Weapons, Talking Peace, by Herbert York, Basic Books, 1987

'Man of Space', *Time*, February 17, 1958

'Memorandum for the President from Administrator of NASA and Secretary of Defense regarding responsibility and organization for certain space activities, October 21, 1959', Dwight D. Eisenhower Presidential Library, Museum and Boyhood Home

'The Missile Design Bureaux and Soviet Manned Space Policy: 1953-1970', a thesis by William Barry, Oxford University, 1996

'Nazi Whitewash in 1940s Charged', by Ralph Blumenthal, *The New York Times*, March 11, 1985

Operation Paperclip, by Annie Jacobsen, Little, Brown & Company, 2014

Recoverable Booster Proposals: A Future That Never Was … , edited by Tony Landis, US Air Force Materiel Command, 2021

'Recovery of Dog Called Possible', *The New York Times*, November 5, 1957

'The Rise and Fall of Dyna-Soar: A History of Air Force Hypersonic R&D, 1944-1963' by Roy Franklin Houchin II, Auburn University, 1995

A Rocket Drive for Long-Range Bombers, by Eugen Sänger and Irene Bredt, Deutsche Luftfahrtforschung, 1944

Rockets and People: Creating a Rocket Industry, Volume II, by Boris Chertok, NASA History Division, 2006

'Sagan breached security by revealing US work on a lunar bomb project', correspondence by Leonard Reiffel, *Nature*, May 2000

'Satellite New Threat to US, Russell Says. Called Disastrous Blow to Prestige', *Chicago Daily Tribune*, October 6, 1957

'Screen: About von Braun: *I Aim for the Stars'*, by Bosley Crowther, *The New York Times*, October 20, 1960

'Space Superiority': Wernher von Braun's campaign for a nuclear-armed space station, 1946–1956, by Michael Neufeld, in *Space Policy* 22, 2006

'Special Mission V-2: Col. Holger Toftoy and the Operation That Brought German Rocket Technology to America,' by Dwight Jon Zimmerman, Defense Media Network, April 26, 2016

Spies in Space: Reflections on National Reconnaissance and the Manned Orbiting Laboratory, by Courtney V.K. Homer, The Center for the Study of National Reconnaissance, May 2019

'A Study of Lunar Research Flights, Vol. I', by Leonard Reiffel *et al*, Research Directorate, Air Force Special Weapons Center, Air Research and Development Command, Kirtland Air Force Base, New Mexico, 1959

'Survey of Development of Liquid Rockets in Germany and their Future Prospects', by Wernher von Braun, *Journal of the British Interplanetary Society*, March 1951

Ten Presidents and NASA, by John Logsdon: https://www.nasa.gov/50th/50th_magazine/10presidents.html

'Thomson Rolling Von Braun Biopic Though Opposition High in Britain', by Hank Werba, *Variety*, August 1959

'US serviceman helped liberate Nazi concentration camp', by Mike Richman, US Department of Veterans Affairs blog: https://blogs.va.gov/VAntage/72294/u-s-serviceman-helped-liberate-nazi-concentration-camp/

'An unintended consequence of the IGY: Eisenhower, Sputnik, the Founding of NASA', by Roger Launius, *Acta Astronautica*, November 2009

'Vanguard Rocket Burns on Beach; Failure to Launch Test Satellite Assailed as Blow to US Prestige', by Milton Bracker, *The New York Times*, December 7, 1957

Von Braun: Dreamer of Space, Engineer of War, by Michael Neufeld, Vintage Books, 2007

'Von Braun Will Leave NASA For Job in Aerospace Industry', by Harold M. Schmeck, Jr., *The New York Times,* May 27, 1972

'Vostok lands successfully', by Anatoly Zak, Russian Space Web: http://www.russianspaceweb.com/vostok1_landing.html

WW2 People's War archive, contribution by Eric Taylor, November 13, 2003: https://www.bbc.co.uk/history/ww2peopleswar/stories/71/a2037971.shtml

'We Built the Saturn V', by Mark Betancourt, *Air & Space*, October 2017

'Wernher von Braun, the SS and Concentration Camp Labor: Questions of Moral, Political and Criminal Responsibility', by Michael Neufeld, in *German Studies Review*, February 2002

'Who got America to the moon? An unlikely collaboration of Jewish and former Nazi scientists and engineers', by Ralph Vartabedian, *The Los Angeles Times*, March 1, 2020

'Why I Chose America', by Wernher von Braun, *The American Magazine*, July 1952

Index

About the Author

DAVID AXE is a journalist, historian, filmmaker and former war correspondent. A prolific contributor to *Forbes, The Daily Beast, Rolling Stone* and many other publications, David has also directed independent movies including the horror-comedy *Bae Wolf*. David has written and edited several nonfiction books, most recently the graphic novel war memoir *Machete Squad* and the Pen & Sword book *Drone War: Vietnam*.